Right Now: The Book of Odd

A New Perspective for This Reality

By Todd R. LeBlanc

Copyright © 2023 by Todd R. LeBlanc

All rights reserved.
No part of this publication may be reproduced, stored in a retrieval system, or transmitted in any form or by any means, electronic, mechanical, photocopying, recording, scanning, or otherwise, without the prior written permission of the author. For more information, address toddrleblanc@gmail.com

This publication is designed to provide accurate and authoritative information in regards to the subject matter covered. While the author has used their best efforts in preparing this book, they make no representations or warranties with respect to the accuracy or completeness of the contents of this book and specifically disclaim any implied warranties of merchantability or fitness for a particular purpose. The advice and strategies contained herein may not be suitable for your situation. You should consult with a professional when appropriate.

First paperback edition December 2023

Editor: Catherine Nikkel
Book Cover & Interior Designer: Julia Baxter
Publishing Consultant: Makini Smith

ISBN: 978-1-7382319-0-4 *Paperback*
ISBN: 978-1-7382319-1-1 *E-book*

TABLE OF CONTENTS

Prologue i
Introduction ii
Tests...11
Is our Reality Self-Emergent? 22
The Moon 49
Is our Existence a Means for the Universe To Comprehend Itself?..60
5 Forces..77
The Small: the Enigma of Quantum Mechanics 90
My First Hint at a Bigger World 102
The Living...122
The Big, Classical Physics and General Relativity..................129
Breadcrumbs 137
And, Oh, I Forgot, What the Heck Is Time? 146
Why Are We Here? 166
Sorcerer's Sandbox...192
Quotes 218
Odd Quotes 228
Glossary...232
About the Author 263
In Search of the Unknown 264

Prologue

Dear Scarlett,
Welcome to this vast and wondrous world! As you embark on your journey, I encourage you to embrace life fully and hold on tightly. The beauty of this reality is highly personal, and it increases logarithmically with your ability to appreciate it. My hope is that you will always keep seeking the wonders of the world, no matter how big or small.

Remember that the beauty of this world is not always easy to find, but it is always worth the effort. Keep delving deep and exploring new perspectives, and you will discover a wealth of beauty that will inspire and sustain you throughout your journey.

I wish you all the best on your adventure.
With love,
Pippi

To my beloved Aurora Dawn,
Dawn holds significant symbolism in various mythologies around the world. In Greek mythology, she is personified as Eos, the goddess of the dawn. Eos, who shares your radiance, was often depicted as a stunning young woman, gracefully announcing the arrival of a new day.

In Roman mythology, Aurora, who bears a resemblance to your enchanting presence, represents the Roman counterpart of Eos. She is associated with the breaking of dawn and the dispersal of darkness. Aurora was often portrayed as a radiant goddess, riding a celestial chariot across the sky, heralding the majestic ascent of the Sun.

Dawn is commonly associated with renewal, hope, and the beautiful cycle of life. You are my inspiration. Thank you.

INTRODUCTION

The forthcoming text is a profound journey into philosophy and creative expression, aiming to offer a new perspective for decoding the intricate tapestry of our existence. It aspires to unravel the mysteries surrounding us, providing insightful guidance as we navigate the complexities of our shared reality.

We live in the now.

The past is a memory of things that were: pictures in albums, songs we heard as kids, people who loved us and are now gone. Everything that ever happened in the past can only be seen through the mind's eye.

The future is an idea of things we want to be: hopes, dreams, prayers, and goodwill, children who call us theirs. Everything that will ever happen tomorrow can only be seen through the mind's eye.

All we have is now. Do not waste it.

Everything that has ever happened and everything that will happen is converging on the now. Infinity expanding in both directions from a single point we call today...

Why I wrote this book?

The answer to this simple question is... I did not start out writing a book. I started out trying to improve.

At the outset of the coronavirus pandemic, our collective worry was palpable. The looming threat of this unseen virus had the potential to wreak havoc on our lives, and indeed, it did. Hardships emerged on multiple fronts, as we grappled with the unforeseen challenges it presented.

For countless individuals, job security became a fragile concept as layoffs rippled through various industries. The economic collapse that followed cast a shadow of uncertainty over financial stability, leaving many in a state of financial distress.

Working from home became the new norm for a vast portion of the workforce, blurring the lines between professional and personal life. The transition to remote work brought its own set of challenges, from adapting to virtual meetings and the isolation of home offices to juggling household responsibilities alongside professional duties.

Meanwhile, sickness struck homes and communities, leaving a trail of illness and loss in its wake. Families faced the heart-wrenching reality of dealing with the virus's impact on their loved ones' health.

Amidst all these disruptions, another unexpected consequence unfolded — crazy house prices. The real estate market, influenced by shifting priorities and remote work trends, witnessed a surge in demand, driving property values to unprecedented levels.

Yet, despite these tumultuous changes and uncertainties, life goes on. Humans, with their remarkable adaptability and resilience, found ways to adjust and endure. It's a testament to our capacity to navigate through adversity and continue moving forward, even in the face of unprecedented challenges.

I decided to venture into management courses in an effort to enhance my skills. Despite my enduring interest in learning, my dyslexia and spelling challenges have always left me feeling unsure and self-conscious about pursuing higher formal accomplishments. With the pandemic enforcing lockdowns, I saw an opportunity to work on myself. Numerous online courses were available, conveniently fitting into my schedule, mainly during weekends and evenings. This adjustment didn't disrupt my routine significantly, given the surplus of time I had then.

Among the courses I undertook, a few held my fascination, while others wore a more business-oriented demeanour. I discovered myself gravitating toward courses that sparked inquiries, finding myself less captivated by those with a business focus. Little did I know my affinity for philosophy was on the cusp of emergence.

A year later, I had successfully earned a leadership certificate from a prestigious university. However, what carried even greater importance was the profound insight I had gained into my authentic learning preferences and my capacity to excel with the support of my spell-checking computer. Notably, during this period, my confidence and curiosity experienced exponential growth.

One of my final courses explored the DNA of top performers, taught by a popular and extroverted instructor who assigned copious homework.

For the ultimate evaluation, we were tasked with choosing an essay topic from a shortlist that encapsulated the course's essence. I selected a topic involving a time-travelling future self narrating to the present self, focusing on the DNA of high achievers. Instead of concentrating on mundane subjects like stock predictions or lottery numbers, my story's focus was the nature of reality. This storyline appealed to me due to my deep curiosity for time travel tales and the fascination I held for DNA.

I started with: Why are we here?

Our emergent reality is built on a foundation of fundamental laws and building blocks that have expanded to form the small world we call home. Our home has, in turn, seen many miraculous emergent events, namely the emergence of DNA and the emergence of the conscious man.

It was the start of my journey.

In the realm of education, feedback plays a significant role in shaping a student's learning journey. While both positive and negative feedback has its place, the power of positive feedback often stands out as a motivational force capable of inspiring students to excel and reach their fullest potential. My own experience serves as a testament to this.

Engaging with the subject matter of my essay, I found the words flowing effortlessly from me as I carefully crafted my thoughts and ideas onto the page. The process was both exhilarating and fulfilling as I delved deeper into the topic at hand. The final result was a piece of work that I was genuinely proud of - a product of my dedication and passion for the subject.

Upon submission of my essay, I eagerly awaited the instructor's feedback. To my delight, I received a commendable grade and an abundance of praise from my instructor. His enthusiasm and encouragement mirrored my own, and this positive reinforcement left a profound impact on my outlook.

The potency of positive feedback became evident. It was not just about receiving accolades; it was about the sense of validation and affirmation it provided. It served as a clear indication that my efforts had not gone unnoticed and that my work was appreciated. This positive reinforcement not only boosted my confidence but also motivated me to strive for excellence in all my academic pursuits.

Such was my optimism that I felt compelled to revise my essay, this time focusing solely on the intellectual pursuit, with no strings attached. The positive feedback sparked a desire within me to improve further, refine my work, and explore the subject matter from new angles. It was not about seeking external validation but rather an intrinsic drive to become better.

The power of positive feedback is undeniable. It has the potential to transform a student's educational experience by instilling confidence, fostering enthusiasm, and inspiring continuous improvement. It reminds us that acknowledgment and encouragement can be powerful tools for motivation and growth. As learners and educators, we should recognize and harness the potential of positive feedback to cultivate a culture of excellence and lifelong learning.

And three long years later, here we are.

The Power of Optimism

Optimism is the belief that our world can be improved with effort. This notion carries immense significance. It means that, despite adversity, we hold the conviction that our actions, determination, and creativity can lead to positive change.

Optimism is a force that has shaped history. It fueled movements like the Civil Rights Movement, led by individuals who believed in a better future and inspired others to join their cause. It propelled inventors like Thomas Edison, who saw countless failures as steps toward success. In science and environmental activism, optimism is the driving force behind breakthroughs and solutions.

Yet, optimism is not passive; it requires action. Believing in improvement is the first step. Taking determined steps, even in the face of adversity, is what transforms optimism into reality.

In essence, optimism is a catalyst for change, pushing us to strive for a brighter world through our collective and individual efforts.

The following is an expression of my emergent educational journey exploring fundamental questions. Ultimately, my journey led me to a unique destination that I called:

The Book of Odd.

Why are we here?

The age-old question of why are we here has been contemplated by philosophers and kings since the dawn of time. However, the answer to this question is elusive and changes depending on the person being asked. Our reality appears to be subjective, but the answer to this question is not simply a matter of personal viewpoint but an emergent property of the universe itself. The depth of this answer is unbalanced with the simplicity of the question, leading many to embrace false, convoluted explanations. Reality is a participatory process where observers and collective consciousness actively shape the universe. The world we perceive is not a passive, objective reality but a subjective, interactive one that arises from the interplay between consciousness and information through the eye of a conscious observer.

We awaken each day to view the world open before us, unaware of how accurate that metaphor is.

The interaction between collective consciousness and shared fractal information is what allows life to exist. The Conscious Anthropic Principle (CAP) proposes that the universe is conscious first and that reality is self-emergent. This theory suggests that the universe is not merely a random collection of particles and energy but rather a contemplative creation that intentionally fosters the development of intelligent life.

Consciousness interacts with shared fractal information to shape the universe, allowing for the finely tuned conditions we observe. In essence, the CAP suggests that the universe is not a mere coincidence but rather a purposeful creation driven by conscious intentionality and constructed with shared patterns of fractal information.

"The Book of Odd"
There are two key points I wish to communicate:

Strive to become the finest version of yourself. Pose profound questions, and sustain a commitment to learning, growth, and the pursuit of excellence.

Share your finest self, and in doing so, cast a positive influence on the world by actively working to enhance it.

Many times, I'll jot down details to aid my thinking process, and it's important for you to know that this work remains a work in progress. I don't possess all the answers, and I'm not going to insist that this represents the definitive truth.

So, the fine details I write down may not be of utmost importance to me. My primary goal is to convey a message. The message I'm endeavouring to convey is this: ask big questions, educate yourself, ponder, contemplate what is happening all around us and find your own questions.

Warning: Some assembly required.

As the wise Carl Sagan once emphasized, "extraordinary claims require extraordinary evidence."

The scientific method provides a meticulously designed roadmap for inquisitive minds on a quest to uncover the world's enigmas. These curious thinkers assume the roles of scientific detectives, methodically dissecting the clues nature offers. So, let's embark on this journey into the inner workings of the method and embrace the wondrous mysteries hidden in plain sight.

Step 1: Observation
Our journey commences with meticulous observation.

Step 2: Questions
Building upon their observations, curious minds begin to pose questions.

"I have an affinity for questions and enigmas," as questions serve as the gateway to knowledge. Each question ignites curiosity and propels the scientific journey forward.

Step 3: Proposing Testable Explanations

With questions in mind, these inquiring minds propose testable explanations or predictions. These hypotheses are educated guesses, their best efforts to address the questions raised. Tests act as stepping stones on the path to discovery.

I've outlined a series of extraordinary tests, followed by four unusual questions that have led me to this odd perspective. As we embark on this journey of discovery, it becomes evident that it's not for the faint-hearted; it's a quest to unravel the intricacies of reality. Our aim is to comprehend how everything emerges from the very fabric of the cosmos. This endeavour is undeniably ambitious, yet it holds the promise of revealing profound truths.

TESTS

1. This reality seems to have at least two fundamental features. These features affect everything and have no force carriers. Emergence and Gravity.

   Imagine emergence as a fundamental aspect, akin to the familiar gravity. Emergence is the process of complex properties arising from simpler components; I am trying to suggest emergence is a fundamental aspect of how the universe operates. I also suggest that emergence and gravity are two sides of the same coin and work in tandem.

   We observe emergence in the formation of subatomic particles, the twisting of proteins, or the behaviour of ecosystems in response to environmental changes.
   For example, DNA should spontaneously emerge given the opportunity.[1]

   Gravity is not a force. It is a fundamental aspect of this reality, like emergence. On large scales, gravity behaves as we would expect it to behave. On smaller scales, emergence is the dominating influence on how reality behaves. Gravity is not quantum, and it has no force carriers. Gravity is pixelated as information, but so is everything else. Gravity describes how the universe behaves in the presence of mass.

2. We may be alone as the only type of life form in the universe, which statistically seems improbable in a random universe and could suggest causality. The exploration of Mars and other nearby planets in our solar system may provide insights into life in our galaxy.[2]

---

[1] DNA will be found almost everywhere, even in purposely sterilized areas leading to incorrect contamination claims.

[2] Life will be found elsewhere but will be from the same tree of life, shared ancestor or same DNA (RNA).

3. Emergent life theory: The Emergent Life Theory proposes that all life forms have a shared ancestry that emerged from shared information. According to this theory, all extraterrestrial civilizations are currently at a similar stage of advancement.[3]

4. The many fundamental constants of our reality are all derivative products of a single force-like aspect, and their precise values are the values needed to support conscious life and us. An Anthropic viewpoint will shed light on the inner workings of our reality.

5. Our universe only has two dimensions, consciousness and information. We exist in a perceived matrix of reality with length, width, height, and time as an emergent construct of consciousness built with fractal information.

6. Time is what we observe as the progression of events from the past, through the present, and into the future. Time is not a dimension. Time does not exist in the conventional sense; our experience of time only exists in the biological mind of a conscious observer, past as memory, the present as observation, and future as extrapolated experience. Causality exists in the movement of the "Now" to the next "Now" and is part of the emergent construct that is our reality and a product of the expansion of the universe.

7. Our reality appears to be composed of a vast array of tiny fractal pockets that contain lists of information and act according to a simple set of rules without force carriers. These pockets act in concert and hold all the necessary information for the present moment. The information in these pockets is fractal and follows a universal blueprint to determine possible outcomes. Energy and matter exist as information within these pockets and consist of various

---

[3] The ages of life and advanced civilizations in our galaxy should be universal, offering a possible solution to the Fermi paradox.).

ingredients, with mass being just one of them. The format of these ingredients is defined within the fractal pocket. I suspect our perceived three-dimensional realm interacts with a single (infinite) dimension of information.

8. This reality is stored as information within its system of fractal pockets; as the universe expands and causality moves forward, pixelation and blurring occur. Pixelation and quantization are similar in appearance but not exactly the same.

9. Information stored in the fabric of our reality has a small finite mass. Learning how the universe expands will grant us insight into the mysteries of gravity and time (empty space is still information, and large amounts of it should have measurable mass). Dark matter and dark energy only exist if they serve an anthropic purpose.

10. Computers imitating thought will not be conscious, no matter how smart or complex. Conscious artificial intelligence will only be possible by allowing networks to be grown (like crystals or cells) and allowing emergence.[4]

11. The primary function of the human brain is to support our reality. A better understanding of the brain will reveal an excessive overpowering and a metaphysical primary purpose. What we see as superconductivity and quantum entanglement are part of the active mechanism and driver of our biological brain. These phenomena are required to support our purpose and that purpose has an anthropic origin.

12. The strangeness of the quantum world will be seen to have an anthropic purpose. As we delve deeper, we may discover

---

[4] AI could enable the emergence of consciousness if it is equipped with the capacity to observe, mirroring the role of an observer in quantum physics, and provided with adequate resources for this purpose.

that everything has an anthropic purpose. Once viewed as spooky and mysterious, superconductivity and quantum entanglement will be observed in their biological equivalent in essential everyday processes like photosynthesis, brain function, and information sharing. The true nature of these phenomena will be unveiled even at normal biological temperatures, reigniting our sense of wonder and opening new frontiers of understanding.

13. At the most fundamental level, the interconnectedness of all things in the universe reflects the concept of information sharing, with even the smallest components of matter and energy being linked. This essential aspect is crucial to comprehending the workings of our reality, as evidenced by the theory of Morphic Resonance.

    Conscious thought, configurations of how patterns emerge, and vast amounts of data in the form of memories are all stored in the fabric of reality. Qualia, which are the subjective experiences or qualities of conscious perception, such as the way we experience the colour red or the taste of chocolate, suggest that universal fundamental information underlies our subjective experiences.

    Moreover, I believe a breakthrough in fractal pocket memory in data storage should be expected to revolutionize the field. Similarly, fractal pocket communication could facilitate unprecedented knowledge sharing among humanity, enabling us to listen to the stars and better understand ourselves and the universe.

14. The phenomena of group mind, collective memory, placebos, déjà vu, and the Mandela effect are examples of the true nature of our reality. These experiences suggest that our reality is not simply a collection of individual experiences but rather a shared experience that is influenced by the collective consciousness. There is evidence available

that supports these phenomena, which can help us better understand the interconnectedness of all things in the universe.

Questions

In this grand adventure of seeking answers, questions serve as our compass, faithfully pointing the way to knowledge. They act as our trusted guide, kindling the spark of curiosity and steering our scientific journey forward with unwavering purpose and determination. These are the four unusual questions that initiated my quest:

1. Is our reality self-emergent?

> *"Every problem emerges from the false belief we are separate from one another, and every answer emerges from the realization we are not."*
> — *Marianne Williamson*

I believe that emergence constitutes a fundamental aspect of our reality and warrants observation across all scales. Emergence is akin to an elemental feature rather than a force, roughly approximating gravity in nature. This phenomenon manifests at various scales, from the birth of photons to the formation of crystals, or from the captivating movement of fish schools to the intricate ecosystems of grand proportions.

Could emergence, then, be likened to a realization, a pivotal puzzle piece within the present moment of Right Now?

2. Is our existence a means for the universe to comprehend itself?

"We are a way for the universe to know itself. Some part of our being knows this is where we came from. We long to return. And we can, because the cosmos is also within us. We're made of star stuff."
— *Carl Sagan*

Does the concept of Universal Consciousness interacting with Fractal Information allow reality to emerge?

Fractals, repeating patterns at varying scales, are ubiquitous in nature and human experiences, connecting us to the larger cosmic tapestry. These patterns are infused with significance by the universal consciousness — a fundamental essence unifying all existence. This consciousness, akin to a life force, interconnects every aspect of the universe, enabling perception, thought, and feeling. It prompts profound questions about whether Universal Consciousness and Fractal Information are the underpinnings of reality, giving rise to the emergence of space, matter, energy, gravity, time, and life itself, inviting exploration into our deep connection with the cosmos.

And could the evolution of all things be recorded by shared fractal information, forming a blueprint for existence?

3. And, oh, I forgot, what the heck is time?

"When we listen to a hymn, the meaning of a sound is given by the ones that come before and after it. Music can occur only in time, but if we are always in the present moment, how is it possible to hear it? It is possible because our consciousness is based on memory and on anticipation. A hymn, a song, is in some way present in our minds in a unified form, held together by something…by that which we take time to be. And hence this is what time is: it is entirely in the present, in our minds, as memory and as anticipation."
— *Augustine*

Could the Now be described as the dynamic, ever-changing segment of the universe?

Right Now: The "physical" universe and all life, only "physically" exist in the Now. Past frames of existence exist only as information and future events as possibilities.

The present moment defines the path that has led us to where we are now. Even the most improbable events of the past were necessary to arrive at this moment and form. The emergent nature of history, evolution, cosmic events, and the underlying rules of the universe are all based on the present moment.

Thus, could the present moment be an extraordinary factor that steers the course of both the past and future?

And Time, what is our relationship with time?

4. Why are we here?

"The problem, often not discovered until late in life, is that when you look for things in life like love, meaning, motivation, it implies they are sitting behind a tree or under a rock. The most successful people in life recognize that in life they create their own love, they manufacture their own meaning, they generate their own motivation. For me, I am driven by two main philosophies, and know more today about the world than I knew yesterday. And lessen the suffering of others. You'd be surprised how far that gets you."
— *Neil Degrasse Tyson*

Why are we here?

I always come back to this question. Essentially, it's a matter of purpose. I genuinely hold the belief that the elusive answer to this question resides within us, and it has always been there.

We are the avatars of consciousness and manifestations of information.

What is our relationship with the present moment?

We seem to be defined by our many limitations. Our many limits in our reality force a concentration of complexity, and in turn, give us form.

Could it be that by embracing our imperfections, we allow the universe to comprehend the concept of love?

Back to the beginning.

Why are we here?

Essentially, this odd book is a blend of my thoughts interwoven with facts, offering a glimpse into my perception of this peculiar world. My goal is to make it accessible because I genuinely aim to nurture curiosity and contemplation. I sincerely hope that readers will come away from this book with a better understanding of the mysteries surrounding time, DNA, gravity, consciousness, and information. They may not fully grasp every aspect, but I hope they will gain a deeper insight into these questions, recognizing them as clues. Notice their absence of understanding in this astounding enigma that surrounds us and move to action to enlighten themselves.

They may not entirely comprehend the intricacies of all aspects of reality, but at the very least, I aspire to encourage them to contemplate it, to acknowledge that there's something unusual about this place, something strangely familiar and worth questioning, and in response, to ask their own follow-up questions.

Moreover, I'd like them to ponder the concept of emergence, to question the enigma of the passing moment.

Are we, in fact, residing in this odd, movie-like reality?

When we watch a movie, think of it like this: there's a long strip of film, this strip of film is like a long ribbon, and we shine light through it, which makes a picture appear on the big screen. Now, here's the fascinating part, even though only one piece of the picture is shown at any given moment, it happens so quickly that it looks like the action in the movie is smooth and continuous.

But guess what? Our real world works in a similar way. It's like we're experiencing life one tiny piece at a time, just like one frame in a movie. And that's what our reality is — it's like an unfolding picture or experience, with each moment revealing itself one piece at a time.

Furthermore, it's not only occurring one frame at a time, it's also happening one individual at a time. My perception of reality and how I see it moving forward is highly personal and unique to me. You'll likely have a very similar experience, but it's uniquely yours. And while we can't directly share it, we can describe it, talk about it, and share certain things. Certain fractal pieces of information, like the colour red, the taste of chocolate, the scent of vanilla — are shared experiences embedded in the universe's fractal blueprint, the universal cookbook of information.

So, here's the main thing I want to say with this book: it's about asking questions, thinking expansively, refusing to be satisfied with small questions or small answers, and it's also about the journey of self-improvement through putting in hard work and effort. The central message of this book is quite short, only about three or four pages, but it emphasizes these important ideas, including the idea that we can become better by working hard on ourselves.

Our reality, this absurd reality, doesn't make complete sense. So, if someone insists that it all makes perfect sense, that's a sign that you're not asking questions on a grand enough scale, you're not equipping yourself with enough information and knowledge to shield yourself from misleading small truths.

That's the core message I aim to convey.

Hey, I don't know, and that's the perspective I'm offering (I don't know, and neither does anyone else).

Returning to the analogy of the movie projector, I believe this encapsulates our reality, illuminating why these aspects hold significance. Consciousness and information converge to bring about emergence — everything is information. Consciousness constitutes observation; you understand how it all intertwines.

We, as sentient beings, enable time's existence because time solely manifests within us. Through our existence, we serve as the instrument to generate causality, permitting time to progress, ushering in the emergent event of the present moment. We are the avatars of consciousness and manifestations of information. And that, in and of itself, holds profound wonder.

Moreover, we possess the capability to exert influence over the ensuing "Now," and this is the message I passionately desire to convey. We retain the power to shape tomorrow through present-day performance, moulding the upcoming "Now." This power encompasses both our capacity and our purpose. We possess the potential to enhance circumstances.

By collectively uniting and working in concert, provided we align ourselves in the correct direction, we can achieve remarkable outcomes. This is not mere idealism; I genuinely believe this to be the underlying mechanism. We can approach this from various angles, whether through discussions surrounding religion, morality, or the importance of kindness toward others. Nevertheless, it also harmonizes with this theory — that through conscientious efforts today, we can construct a brighter tomorrow.

IS OUR REALITY SELF-EMERGENT?

Question 1

The intricate web of existence implies a more profound interconnectedness, a cosmic puzzle in which each fragment represents information yearning to be unveiled.

But first...

What is a question?

Questions, like the delicate strands of a spider's web, weave through the fabric of our curious minds, connecting the known to the unknown. They are the keys that unlock the doors of understanding, prompting us to probe, ponder, and explore.

But what exactly is a question?

Beyond its mere linguistic construct, a question embodies the essence of curiosity and the pursuit of knowledge.

At its core, a question is a cognitive tool that enables us to seek information, clarify uncertainties, and stimulate thought. It is a quest for insight. Questions might often be seen as simple as "What time is it?" or as profound as "What is the meaning of life?" but the mechanism through which we engage the question and how it is asked can encourage bold thought or shut down dissent.

The art of questioning is a fundamental aspect of human nature. From our earliest moments of existence, we instinctively inquire about our surroundings, driven by an insatiable curiosity.

Children, with their unbounded wonder, are renowned for their incessant stream of questions, each serving as a portal to discover the world. This pursuit continues into adulthood, albeit often with a more subdued intent. We ask questions to solve problems, to deepen our understanding, and to challenge the status quo.

Questions serve various purposes. They can be informative and aimed at acquiring facts or data. They can be exploratory, encouraging us to delve into uncharted territories of thought.

Questions can also be analytical, prodding us to dissect complex concepts into comprehensible fragments. Moreover, they can be rhetorical, designed not to elicit answers but to provoke thought or convey a point.

Asking questions also nurtures a reciprocal exchange. It establishes a bridge between the inquirer and the responder, generating a shared exploration space. When a question is posed, it invites others to contribute their perspectives, experiences, and insights, fostering a collaborative journey toward deeper understanding.

However, the significance of a question extends beyond its practical function. It embodies a yearning for growth and enlightenment. A question signals our acknowledgment of the vast expanse of knowledge that eludes us and our willingness to embark on the quest to bridge that gap. It symbolizes humility in the face of the unknown and resilience in the face of uncertainty.

A question is more than an arrangement of words, it is a vessel of curiosity, an instrument of exploration, and a catalyst for growth. It propels us to stretch the boundaries of our understanding and embrace the infinite possibilities that lie ahead. In the art of questioning, we find answers and the beauty of the journey itself.

"I am not a genius, I am just curious. I ask many questions, and when the answer is simple, then God is answering." — Albert Einstein

I believe we are self-emergent and that emergence is a fundamental aspect of this reality and should be observed on all scales.

That leads to the question:
But what is emergent?

Webster's Dictionary definition of emergent:
Arising unexpectedly.

Wikipedia states that emergence occurs when an entity is observed to have properties that its parts do not have on their own, properties or behaviours that emerge only when the parts interact in a wider whole.

I prefer: Emergence is the act of becoming known or coming into view.

Emergence refers to perplexing manifestations of self-organization that give rise to a sudden injection of complex order. It is fascinating to observe ants assembling themselves into structures, birds gathering in flocks called murmurations, or planets like Saturn having hexagon-shaped storms. These extraordinarily complex phenomena occur spontaneously almost everywhere, raising questions about their origin.

Many things in nature self-organize with just a few simple rules, giving rise to higher orders of complexity and beauty. These new assemblies can have entirely new properties and behaviours that "emerge" from humble beginnings. Remarkably, no one actively controls the process, and the simple individual components never give hints of their exotic destiny. In this reality, the whole is truly greater than the sum of its parts.

Our emergent reality is built on a foundation of fundamental laws and building blocks that have expanded to form the small world we call home. Our home has witnessed many miraculous emergent events where complex behaviours arise spontaneously from relatively simple elements.

Here are a few examples that showcase emergence, and I aspire to convey its fundamental role in this reality.

The Big Bang

"The problem with doing nothing is that you never know when you're finished." — Groucho Marx

One of the most significant theories in modern cosmology is the Big Bang theory, which proposes that the universe originated from a massive explosion. Cosmologists attribute the possibility of the Big Bang to natural quantum fluctuations, allowing for its spontaneous occurrence. In essence, the Big Bang resulted from the universe's inherent instability and the governing laws of nature.

This summary provides a simplified overview of the events within this complex and ongoing field of cosmological research.

- The universe is estimated to be approximately 13.7 billion years old.
- Originating from an extraordinarily dense and energetic single point, the universe underwent rapid expansion.
- During a period known as inflation, the universe experienced an astonishing burst of expansion, growing exponentially and doubling in size at least 90 times.
- The universe is expanding through a force called Hubble expansion, and as it expands, it started to cool.
- As the universe cools, various subatomic particles begin to form.
- Over time, these subatomic particles combine to form simple elements.

- Roughly 400 million years after the Big Bang, vast hydrogen clouds' gravitational attraction resulted in the formation of the first stars.
- The first stars underwent their natural life cycles, culminating in spectacular explosions and creating heavier elements essential for life.
- Approximately 4.5 billion years ago, the Earth was formed.

Something from nothing... well... *everything* from nothing. The Big Bang is our first example of starting small, throwing in a few basic rules, and voilà - something quite amazing emerged.[5]

The Big Bang seems so far away and may seem like an unimaginable distance in time and space. It occurred an exceptionally long time ago, and I don't believe people fully understand or appreciate how we are perched on top of Infinity.

The events of the past, no matter how unlikely, are the route needed to arrive at this form, at this moment, at this place we call today.

The event of the Big Bang is sending us images in the form of information describing events that transpired billions of years ago from a location billions of light-years away. This information is currently being processed, analyzed, observed, and contemplated. We're endeavouring to unravel precisely what it signifies, striving to discover our birth in this place we call home.

A Catholic priest scientist initially put forth the groundbreaking concept of a cosmic egg and faced initial resistance, not because supporting evidence was lacking, but due to its apparent alignment with the biblical notion of creation.

---

[5] This is also an example of our child-like understanding of our unusual universe. The vast majority of minds in the scientific community realize this theory breaks down at the beginning during the Inflationary epoch period. It is generally accepted to be the foundation of our understanding of the universe, even though we suspect it not to be true.

Once, our philosophy, religion, and dreams were intertwined as one. I believe that discovering this truth may restore our faith in science.

Water

H_2O, the "Solvent of Life." The humble water molecule is as basic as it gets. Water is just an oxygen atom sandwiched between two hydrogen atoms. With water, you get a substance with characteristics you would never guess by studying its individual atoms. Water is the only substance on Earth found in all three states of matter: gas (water vapour), liquid, and solid (ice).

Water has many amazing emergent properties that are unique and essential for life. Water is oceans, icebergs, rivers, clouds, ice cubes, raindrops, rainbows, and snowflakes; it is timeless, immortal, and has been recycled daily for millions of years.

From an anthropic perspective, the transparency of water can be attributed to the specific conditions necessary for life to thrive in our purposeful universe, as proposed by the conscious anthropic principle.

Water's emergent properties continue and include cohesion, adhesion, high surface tension, high specific heat, high heat of vaporization, and solvent abilities. Cohesion allows water molecules to attract and bond with each other, forming surface tension that enables droplets to maintain their shape. Adhesion refers to water's ability to adhere to other substances, which is particularly significant when interacting with charged surfaces.

Water possesses the highest specific heat among liquids, meaning it can absorb a considerable amount of heat before its temperature rises. This property helps regulate temperature in ecosystems, preventing sudden fluctuations that could harm organisms. Water's high heat capacity also enables living beings to maintain stable internal body temperatures.

In the early stages of life on Earth, water's high heat capacity allowed for stable temperatures in oceans and other bodies of water, providing a suitable nursery for developing fragile, single-celled organisms. These organisms could only thrive in a more consistent environment, essential for the initial emergence and evolution of life. Water has a specific heat capacity more than double that of many other liquids.

The high heat of vaporization of water requires a significant amount of energy to convert it from liquid to vapour at a constant temperature. This characteristic moderates temperature changes in the environment by absorbing excess heat.

Overall, water's unique properties contribute to a relatively constant temperature, essential for supporting life on Earth.

From an anthropic perspective, water's transparency results from the finely tuned conditions required for life to thrive. The conscious anthropic principle proposes that the universe is not a random assortment of particles and energy but a purposeful creation that fosters the development of intelligent life. Water's transparency is crucial in enabling various life forms, including humans, to perceive and interact with their surroundings.

Transparent substances allow light to pass through, enabling vision and facilitating photosynthesis, both essential processes for many organisms. The specific molecular structure of water, with its symmetrical distribution of electrons resulting from hydrogen-oxygen bonding, leads to relatively low absorption of visible light. This molecular configuration allows most light to pass through water, making it transparent.

If water were not transparent, our world would be drastically different.

Vision, a vital sense for survival and observation, would be impossible. Photosynthetic organisms, such as plants, would struggle to harness sunlight for energy production, disrupting the entire food chain and ecological balance.

Therefore, the transparency of water appears to reflect a purposeful aspect of our universe. It illustrates one of the numerous finely tuned conditions required for the emergence and development of intelligent beings, enabling them to perceive their environment, harness light energy, and engage in Earth's intricate ecosystem.

SIDE QUEST

Have you ever marvelled at the exquisite beauty of frost crystals forming on a windowpane as water freezes? You can experience the captivating process of emergence firsthand. You can witness this remarkable transformation unfold in real time following a simple experiment.

Begin by placing a plastic water bottle in the freezer for approximately 30 minutes. Once chilled, gently position the bottle on the counter and apply a sudden force. As if by magic, the liquid water will swiftly metamorphose into solid ice, a captivating spectacle that unfolds right before your very eyes.

> *"Crystals are the flowers of the mineral kingdom.*
> *The beauty of all reality wrapped up in a promise."*
> — *Odd*

My question to you, my intrepid adventurer: Why do we find crystals beautiful, and what is their source of mesmerizing beauty?

Crystals: The very term "crystal" evokes the concept of emergence and beauty, revealing the true essence of our reality. Crystals represent a distinct category of solids composed of minuscule entities called molecules or atoms, meticulously arranged in an orderly and repetitive pattern known as a lattice. This intricate arrangement arises from the compulsion of molecules or atoms to orient themselves in a specific manner, seeking enhanced stability and reduced energy consumption, resulting in the lowest entropy state.

The formation of this lattice grants crystals their unique and intriguing properties, which vary depending on their composition. For instance, diamonds exhibit exceptional hardness and brilliance, salt crystals possess distinctive shapes and dissolve readily in water, and certain metals can be magnetized.

This process of crystal formation emerges through a physical phenomenon known as spontaneous symmetry breaking. In this process, molecules or atoms deviate from their original symmetry and reconfigure into a new and more stable arrangement, ultimately giving rise to the formation of a crystal.

On my long and often challenging journey towards enlightenment, I've had the privilege of stumbling upon numerous opportunities for learning. One of the most eye-opening experiences was when I took on the role of a production supervisor at an automated solar power plant. Little did I know this endeavour would unveil a world of questions and spark the flames of curiosity within my odd mind.

As I delved into the intricacies of overseeing the operations of this cutting-edge solar power facility, my inquisitive nature frequently led my thoughts astray. I found myself pondering the mysterious qualities of the materials around me. Questions arose like, *"Why is glass transparent, and why does tempering glass demand such enormous amounts of energy?"* It was as though the material world itself was an enigma waiting to be unravelled.

Glass, a substance so commonly used, became a fascinating subject of contemplation. I couldn't help but wonder why certain materials, like water, naturally allow light to pass through, while others remain resolutely opaque. The baffling nature of honey's transparency, a substance that defied conventional expectations, and carbon's dual identity as both transparent and pitch-black ignited my curiosity.

Amidst these reflections, I also found myself captivated by the allure of gemstones. These precious jewels seemed to emanate an inner beauty, as if they held a fragment of the universe's magic within. The luminescence of these gems stirred a sense of wonder, prompting me to explore the intricate forces responsible for their captivating radiance.

However, my musings weren't limited to the materials themselves. Stained glass windows, those exquisite creations of art and architecture, also captured my imagination. The interplay of coloured glass and light had a remarkable ability to transport one's thoughts to a realm that felt nothing short of divine.

But in the midst of all these wonders, one sobering reality persisted: the vast amounts of energy required to temper glass and to manufacture such materials. It was a stark reminder that every advancement in our understanding of the world comes at a cost — a cost we must consider in our pursuit of knowledge and innovation.

My journey towards enlightenment has been a labyrinthine exploration of the mysteries that surround us. It's a voyage where I've encountered a multitude of learning opportunities, each presenting new questions and insights. It's a journey where the ordinary becomes extraordinary, and where the quest for knowledge becomes as essential as the very air we breathe.

In my quest to fathom the transparency of common elements like water or glass, the beauty of gems, and the enchantment of stained glass, I've discovered that the path to enlightenment is not a final destination, it's a lifelong expedition filled with curiosity, wonder, and a relentless desire to unveil the universe's most profound secrets. Amidst this journey, I've also come to appreciate the delicate balance between our quest for knowledge and the energy required to achieve it.

Life

Life: *Are you an emergent phenomenon?* You started out as a single cell. Add a few rules and 75 trillion cells later, and you are reading this question.

But what is life?

There is currently no consensus regarding the definition of life. One popular definition is that organisms are open systems that maintain homeostasis, are composed of cells, have a life cycle, undergo metabolism, grow, adapt to their environment, respond to stimuli, reproduce, and evolve.

I believe life is simply self-replicating information. Some may believe this is too broad a definition and does not account for many things we see life doing. However, I believe having a definition that could describe life at its simplest starting point is necessary. At the beginning.

How did life start?

Let's start small and work our way up.

Deoxyribonucleic acid, commonly known as DNA, is an intricate molecule of remarkable complexity, encompassing all the essential information needed to construct and maintain an organism, including yourself. Biological DNA represents quaternary information, employing a base 4 equation composed of the letters A, C, G, and T, constituting the foundation of life's information repository, the language of life. Within the vast realm of DNA, one can find a wealth of structured data, encompassing order, sequence, and the instructions for assembling the 20 amino acids essential for life, which in turn build proteins.

DNA serves as an extraordinary self-solving puzzle piece, seamlessly fitting into each and every cell within our bodies, intricately constructing us from beginning to end. Our physical characteristics take form through the DNA residing within our cells, granting us shape and purpose. Your DNA is the blueprint of you.

SIDE QUEST

Do you know what happens when a spider is exposed to a high dose of radiation?

Contrary to popular belief, you do not get a mutant-enabling super spider. Instead, you get a very sick spider. DNA is a repository of information, and introducing random errors to this information ultimately destroys it. Noise, in the context of information technology, is an extensively explored field that focuses on understanding and mitigating such disruptions.

It may seem challenging to envision your body as an intricate and vast arrangement of atom-based miniature orbs, but this fascinating concept holds true. However, it may be more accessible to visualize matter arranged in increasing levels of complexity.

For instance, subatomic particles can assemble into atoms, which serve as the building blocks of molecules. These molecules, in turn, organize themselves into more intricate structures like DNA and proteins, which then form the foundation of cells. In their collective organization, cells give rise to tissues, which subsequently combine to form organs. Organs, when integrated, constitute organ systems, ultimately culminating in the creation of entire organisms, including humans like ourselves.

But is Life emergent?

Let us look at what we know and try to deduce a logical conclusion.

- The Earth was formed 4.5 billion years ago.
- Fossils from 3.8 billion years ago reveal primitive life. DNA or RNA.
- DNA is an extraordinarily complex molecule and contains all the information of life.
- 3.8 billion years ago, the Earth cooled to a point where liquid water could exist.
- Life has emergent properties.
- Life was observed on Earth as soon as it was possible. Instantaneously (give or take a million years — which is actually quite short compared to the total time life has been around but much longer than all of human history).
- We only have one tree of life. All life on Earth is descended from this one occurrence, and in the last 3.8 billion years, it has never happened again.

With our limited information, I believe we can state that life emerged spontaneously or was brought to early Earth by an external mechanism (see Test 2: We are alone).

Evolution vs. Emergent Process of Life (EPL)

Evolution is a change in the heritable characteristics of biological populations over successive generations. These characteristics are the gene expressions passed on from parent to offspring during reproduction. Different characteristics tend to exist within any given population as a result of mutation, genetic recombination, and other sources of genetic variation. Evolution occurs when evolutionary processes such as natural selection (including sexual selection) and genetic drift act on this variation, making certain characteristics more common or rare within a population. This process of evolution has given rise to biodiversity at every level of biological organization, including the levels of species, individual organisms, and molecules (Wiki).

I have never been satisfied with this definition. It's not that I believe it is wrong, it's just not complete. My dissatisfaction and interest in the subject were strong enough for me to read *Origin of Species* by Charles Darwin (I strongly suggest reading this for yourself).

Not only did Charles Darwin not coin the phrase *"survival of the fittest"* (the phrase was invented by Herbert Spencer), but he argued against it.

In *On the Origin of Species,* he wrote: "It hardly seems probable that the number of men gifted with such virtues (as bravery and sympathy)... could be increased through natural selection, that is, by the survival of the fittest. Cooperation has been more important than competition in humanity's evolutionary success. Compassion is the reason for both the human race's survival and its ability to continue to thrive as a species."

I was utterly amazed by Darwin's power of observation and his rigid discipline to adhere to the scientific method, even when his observations were against his own worldview.

Allow me to clarify, I readily acknowledge my occasional lapses in discipline, but it's crucial to understand that my approach to this intricate puzzle is firmly grounded in the realms of logic and observation. My journey, which initially began as an endeavour in creative writing and philosophy, was aimed at exploring the intricacies of existence, primarily addressing the first two questions that had piqued my curiosity (more on the first two questions in "The Small").

DNA, with its unparalleled complexity and intricacy, stands as a profound enigma, a puzzle that beckons us to unravel its mysteries. It's more than just a molecule, it's a question mark, a riddle waiting to be deciphered. When I encounter something that doesn't quite fit, something that challenges the boundaries of our understanding, I find myself irresistibly drawn to it.

Life, in all its beauty and complexity, often presents us with clues and mysteries, some as overwhelming as the crashing waves of an ocean, while others are so subtle that they go unnoticed, taken for granted as constants of our existence. In the presence of intellectual giants who have shaped the world with their wisdom, I occasionally feel humbled and unworthy of questioning the established status quo. It's easy to be daunted by the vast reservoir of knowledge that surrounds us.

Yet, this is where the beauty of innocence and a short memory comes into play. I am fortunate to possess a genuine and natural curiosity, unburdened by past apprehensions. With this fresh perspective, I find the courage to question, to challenge, and to seek answers. I see the odd and the unconventional as beautiful clues, beckoning me to delve deeper, to investigate, to unearth, and to learn and grow.

The pursuit of knowledge is an ongoing journey, marked by the questioning of the familiar and the exploration of the unfamiliar. DNA, with its intricate code of life, is an enigma that continues to captivate our collective imagination. It reminds us that the universe is far more complex and awe-inspiring than we can fully grasp. Our humble journey of curiosity is an integral part of unravelling its secrets.

In the grand tapestry of existence, each of us plays a unique role as a seeker of answers, a solver of mysteries, and a contributor to the ever-evolving story of human understanding. The journey is not without its challenges and occasional lapses in discipline. However, as long as we remain rooted in logic, observation, and the beauty of innocent curiosity, we are bound to leave our mark on the world, one question at a time. The pursuit of knowledge, guided by curiosity and humility, is a testament to the power of the human spirit and our insatiable desire to explore the mysteries that surround us.

SIDE QUEST: THE SCIENTIFIC METHOD

Imagine you're like a detective, but instead of solving mysteries, you're figuring out how things work in the natural world. Scientists use a special method called "The Scientific Method" to do this. It's like a step-by-step guide.

Step 1: Observe
First, you start by looking at something in nature, like a plant growing or a ball rolling. You pay very close attention to it.

Step 2: Ask Questions
After watching, you might wonder why something happens the way it does. So, you ask a specific question about it. For example, why do plants grow toward the sunlight?

Step 3: Make an Educated Guess
Next, you make a guess, called a "hypothesis," about why things happen the way they do. It's like an educated prediction. For our plant question, your hypothesis might be that plants grow toward the sunlight because they need it for food.

Step 4: Test Your Guess
Now, it's time to test your guess. You set up experiments, kind of like science tests, to see if your hypothesis is right. You change some things and keep others the same to compare results. For our plant example, you might try growing plants in a darker place, like a basement, to see if they still grow toward the light.

Step 5: Crunching the Numbers
Once you've done your experiments, you collect all the data, which can be information or measurements. You use math and special tools to make sense of it all.

Step 6: Making Sense of It All

Based on your data, you decide whether your guess (hypothesis) was correct. If your plants grew and bent towards a distant window in your basement, your hypothesis might be correct.

Step 7: Sharing What You Found

Lastly, you tell other scientists and regular folks what you discovered. You do this by writing papers, making presentations, and talking about it. This way, others can learn from your work and maybe even test it themselves.

And here's the cool part, this process doesn't stop! Your discoveries can lead to more questions and guesses, starting the detective adventure again. This scientific method helps us learn about the world in a careful and organized way so that we can understand it better. It's like solving a never-ending puzzle, one piece at a time.

Evolution can be a buzzword and a magnet for polarizing thought, so with that in mind, I separate my thoughts about the Emergent Process of Life from Evolution.

"I would rather have questions that can't be answered than answers that can't be questioned." — Richard P. Feynman

In a world of diverse perspectives and ideologies, the term "Evolution" stands as a wall of force, drawing staunch advocates and ardent skeptics into a whirlpool of passionate discourse. Its implications stretch far and wide, reaching into realms of science, philosophy, and even belief systems, often sparking heated debates and entrenched positions. Amidst this tempest, I recognize the potential for polarization and division accompanying the word.

With a mindful awareness of the divisive currents that often accompany discussions of evolution, I intentionally chart a different course. I steer my intellectual ship toward the calm and uncharted waters of the emergent process of life, where the focus shifts from contentious debates to a holistic exploration of life's awe-inspiring journey.

The notion of emergence, like a kaleidoscope of intrigue, invites me to delve into the intricate tapestry of existence itself. It encompasses the delicate interplay of myriad forces, the harmonious convergence of elements, and the extraordinary symphony of interactions that give rise to the rich tapestry of life forms on our planet. While evolution undoubtedly contributes to this symphony, I choose to spotlight the broader narrative of life's emergence.

In cultivating a space where questions take precedence over dogma and curiosity flourishes without predetermined conclusions, I seek to foster an environment of healthy questioning. Through this lens, I engage with the process of emergence not merely as a scientific concept but as a springboard for growth, reflection, and the kindling of intellectual curiosity.

As I distance myself from the contentious landscape often associated with evolution, I embark on an exploration that encourages diverse questions. How do intricate ecosystems come into being? What unseen forces shape the emergence of new species? How does the dance of genetics and environment contribute to the mosaic of life's evolution?
These queries, unfettered by the weight of divisive discourse, serve as catalysts for intellectual expansion and nuanced understanding.

In this pursuit, I aim to cultivate a mindset that embraces the art of inquiry and empowers others to join me on this journey. By channelling our collective energies into embracing the marvels of emergence, unburdened by the polarization that shadows the term "evolution", we open doors to discoveries, insights, and perspectives that enrich our understanding of life's breathtaking complexity.

Without further adieu, we move on to the main event. Let us see where this leads us.

Evolution

Evolution is a change in inheritable information. I like this simplified version. I believe it is correct, but it does not describe the whole story.

The Emergent Process of Life (EPL)

The whole story. So, what is life doing?

- Propagating information
- Changing
- Innovating and exploring emergent outcomes

Propagating Information

That sounds easy enough... to propagate information. What information is being propagated? Well, the information of life. Ok... What is the information of life, and where did it come from?

The information being propagated is mostly the DNA of simple lifeforms, but the propagation of DNA from all lifeforms seems to be a top priority of life on Earth. The total life biomass on Earth is about 550–560 billion tonnes; of that, around 50 billion tonnes is DNA. Roughly 9% of all biomass is the mass of information stored in DNA. For comparison, total human mass makes up roughly 0.01%.

SIDE QUEST

If all the DNA in a single human cell was placed end to end, it would be six feet long. If all the DNA in all the cells in a human body was placed end to end, it would reach the Sun and back 600 times. A single, simple bacterial cell typically contains one circular DNA molecule. The number of atoms in the DNA of a single bacterial cell can vary depending on the size and type of bacterium. On average, the DNA in a bacterial cell may contain around 1.6×10^9 (1.6 billion) atoms. The information of life emerged spontaneously 3.8 billion years ago. Everything (and I do mean everything) else emerged spontaneously 9.7 billion years before that. DNA is a clue, a question, an enigma.

Changing

What is changing?

The inheritable information is changing.

A more interesting question would be, *what is not changing?*

The language of information is not changing. It never has (this is important!).

The inheritable information is changing, but not the language it is written in. Inheritable information changes for many reasons, including natural selection (including sexual selection), genetic drift, environmental factors, and random mutations, but the language in which this information is written never changes.

This leads me to two questions.

What does an inheritable change look like without changing the language it is written in?

and...

Why has the language not changed in the last 3.8 billion years?

Inheritable information change without changing the language it is written in is normally quite mundane. It happens daily, but given time and the right conditions, it can lead to new species and even more radical destinations. It can also change quite quickly in evolutionary terms. Information can be copied and/or transferred, sometimes in multiple loops or logarithmic patterns, resulting in revolutionary evolutionary change.
(Life is constantly changing; even organisms considered 'living fossils,' such as scorpions, alligators, and sharks, continue to evolve. This phenomenon is a subtle example of convergent evolution.)

Why has the language not changed in the last 3.8 billion years?

The short answer is, I do not know.

In the late 2000s, scientists could add new letters to the language of DNA artificially, so it is not impossible. Natural selection (including sexual selection), genetic drift, environmental factors, and random mutations all had time and opportunity to make this change, and it just did not happen.

The language of life is written on DNA's four-letter alphabet, A, C, G, T, and life's 20 amino acids, which is DNA or the cookbook of life. This language seems to be fixed and does not change. The story it tells is constantly being updated, but the language it is written in never changes.

SIDE QUEST
The language of life is written with the four-letter alphabet of DNA, A, C, G, T, and life's 20 amino acids, from the cookbook of life. Life has only used these four letters and these twenty amino acids, and it's never changed.

I don't know what that means, but it's an important clue hidden in the cookbook of life.

I also really hope you appreciate my vulnerability because I honestly don't know, but when I see something that just doesn't make sense, like why has DNA never changed, and why is there only 20 amino acids, not 21, not 25, not 100? And why does that number never change?

I know it is an important clue, and it makes me say, like, what the heck is going on?

It's been that way for billions of years, and it's never changed. And all those people who say that it's just random if you wait long enough, it'll all just happen; it just statistically does not work out. It happened once, and it almost happened at the very beginning.

Life happened; it happened before water, so basically, just before the earth was cool enough to start to have water, it had life. It didn't have to wait millions or billions of years for the right thing to happen; it happened immediately. I don't know what that means, but I know it's important to note that life happened immediately, and its language has never changed. All life on earth is a reflection of that one instance, and it's never happened again.

You know, viruses and mushrooms and all kinds of fungi, birds, animals, and fish are all related, and it's all through this DNA. I know as I said earlier, these are questions, but that's what I started with.

I started with questions. I wrote an essay on the two questions, DNA and time travel, and I expanded that essay from DNA and time travel, which I really didn't understand.

I didn't understand DNA and I didn't understand time travel, but by expanding those two questions, I ended up here. This is where I am now. I know it's an odd story. I don't know why, but that's exactly what happened. And if I can get to this point just by expanding those two questions, I'd like to bring other people with me on this journey, and maybe they could either fill me in or add more questions. It would be nice to have a couple of answers, but you know, maybe questions are enough.

Innovating and Exploring With Emergent Outcomes

Does nature evolve randomly, or does it follow tried and true methods, and does it have predetermined outcomes?

Evolution as a theory makes sense. However, the advancement of life on Earth makes no sense whatsoever. The animals on earth are far too evolved, and life seems to make intuitive leaps in advancement.

"Politics is the art of looking for trouble, finding it everywhere, diagnosing it incorrectly and applying the wrong remedies." — Groucho Marx

Flight appears to have evolved separately four times in history: in insects, pterosaurs, birds, and bats. These four groups of flying animals did not evolve from a single flying ancestor. Instead, they all evolved the ability to fly from separate ancestors that could not fly. *This makes flight a case of convergent evolution, but convergent into what?* I believe flight could not evolve by the conventional ideas of change without purpose.

Why?

- Too fast. The jump from non-flight to flight happens in an instant (in evolutionary terms).
- It would require hundreds, if not thousands, of generations of non-flying creatures with useless wings to

survive and produce offspring with even bigger, more useless wings.
- Modern-day non-flying birds evolved from flying birds. In fact, no non-flying recent ancestor exists from any of the four groups. Example: birds are birds, dinosaurs are dinosaurs, birds are not dinosaurs.
- Gliding is not flying; the adaptations needed to glide do not lend themselves to flight. Gliding is stretching out fur or skin or ribs. Flying is giving up arms. No known flying creature has evolved from a known glider.
- Requires multiple intricate mutations working in unison. Example: hollow bones, warm-blooded and advanced nervous system.
- Pterosaurs... where to begin... giant flying giraffes, were warm-blooded, had hollow bones, had fur-like feathers (before fur or feathers were invented), and no fossilized record of any recent ancestor exists.
- Bats, no fossilized record of any recent ancestor exists.

Once would be hard to believe... But four times?

Biologists estimate the eye has evolved more than 50 times independently in species such as flies, flatworms, molluscs, and vertebrates. The human eye and octopus eye are almost identical (see Test 13, information is shared in our reality).

ODD FACT

We believe the human eye has at least 50 MP of visual resolution, making it a masterpiece of observation. Picture your eye as a precious, one-of-a-kind Stradivarius violin in the grand orchestra of vision, creating beautiful, intricate music with every glance. Meanwhile, your everyday camera, with its 20 megapixels, is like that friend who insists on playing a kazoo during the concert. Your eye, the Stradivarius, effortlessly steals the show, while the camera's kazoo antics add a touch of whimsy to the performance. And don't get me started with eagles...

How did mammary glands develop? The mammary gland itself is a complex organ, but the numerous other advancements needed in conjunction are difficult to imagine. Working glands, support networks, hormones, mother instinct behaviour, and newborn instinct behaviour.

Flowering plants and their insect pollinators appear to have evolved together simultaneously, providing a classic example of coevolution.

Inherited behaviour: Inherited behaviour or instinct is a complete mystery. All inherited behaviour or instinct information is transferred from parent to offspring through some unknown mechanism. This shared information can only be described as a truly miraculous and an emergent event. All the information needed to physically build the lifeform in question, all the "firmware," "software" and "operating system" transferred from one generation to the next. The emergent process needed for life is difficult to grasp and must involve an information transfer system on a grand scale. Perhaps, morphic resonance posits that "memory is inherent in nature" and that "natural systems... inherit a collective memory from all previous things of their kind" (*Morphic Resonance*, Rupert Sheldrake).

This is by no means a complete list of all emergent behaviour that seems to dominate all life on earth; I have just scratched the surface. Other examples are hagfish, large symbiotic biomes, and the mystery of human evolution. The norm of the development of life is innovation and surprise endings. I think we can safely say evolution is a change in inheritable information. I think we can also say that life on Earth is changing in ways that can be called emergent. The biggest questions in this debate are the questions centred on purpose and have nothing to do with evolution.

So, what is the purpose of life?

To convert inert matter to information?
To give that information eyes?
To allow the birth of humankind?
Or is it simply impossible to tell from our perspective?

Are some outcomes more likely than others because of some unknown rules within this reality?

Are we operating in deterministic loops in otherwise non-deterministic space?

Is it possible that we are part of one emergent event that is not yet complete, and until complete, the rules cannot be changed?

THE MOON

The Moon: (Honorable mention)
Not emergent, but synchronous. The Moon is obviously not impossible, but the Moon is ridiculously improbable.

> *"I like to think the moon is there even if I am not looking at it."*
> —Albert Einstein talking about the strangeness of quantum physics.

- The Moon resulted from a planet-to-planet collision 4.5 billion years ago. Earth and a small Mars-sized planet named Theia.
- Theia is believed to have lost its core during this collision, and it is thought that Theia's and Earth's core merged. This unusual origin story is thought to allow our Earth's emergent life-enabling and protective personality. The Moon is what is left of Theia without its core.
- Hundreds of millions of years ago, the Moon was too close for a perfect solar eclipse. 600 million years from now, the Moon will have moved too far away to have a solar eclipse. The solar eclipse is a temporary phenomenon. In order to have a solar eclipse, an exceedingly large number of very unlikely parameters need to be met. The Moon is 400 times smaller than the Sun, and it's much closer to Earth than the Sun. This relative size and distance relationship makes the Moon and Sun appear roughly the same size in the sky when viewed from Earth. All this is happening in our Sun's "Goldilocks zone" and is occurring now in the brief sliver of history we call today.
- The moon orbits the Earth once every 27.322 days. It also takes approximately 27 days for the moon to rotate once on its axis. As a result, the moon does not seem to be spinning but appears to observers from Earth to be keeping almost perfectly still. This is called synchronous rotation.

What are the chances that this very improbable event is happening now? Is it an important clue?

And if it is a clue, a clue to what Question?

The Cosmic Comedy, Pondering the Moon

When I lift my gaze to the night sky and catch sight of the Moon, radiant and enigmatic, an intriguing thought brushes against my odd mind. It's almost as though the universe is tuning in, curiously watching the spectacle of my existence. This feeling of fascination and wonder reminds me of the quirky notion known as the "Truman Show" — that syndrome where people think they're the stars of an elaborate reality TV show. It's like life decided to pull off a cosmic prank.

The Truman Show delusion, or Truman syndrome, captures the essence of folks who believe their lives are the central act in a meticulously scripted show. They envision themselves under a continuous spotlight, with every action and thought catalogued for some invisible audience. While rooted in a bit of comedy, this captivating concept taps into our intrinsic truth; we are part of a larger narrative and are actively building tomorrow.

Strangely enough, this peculiar delusion actually highlights a universal aspiration we all share — the pursuit of a meaningful role within the vast universe. And that's where the Moon comes in, not as a camera but as a cosmic mirror reflecting our fascination. As I contemplate the Moon, I'm swept into a saga that's played out over billions of years — a tale of celestial collisions, intricate orbits, and a dance of gravitational forces that could rival any reality show plot.

But let's not get too caught up in the cosmic drama. What matters even more is that I view my life, along with all life on this peculiar planet, as an extraordinary emergent event — a result of intricate interactions, delicate balances, and a series of unlikely plot twists. From the cosmic recipe that birthed our planet to the intricacies of life's evolution, we're all players in a wonderfully captivating and slightly absurd narrative.

I hope I always marvel at the Moon's place in the cosmos, its synchronized spin with Earth, always smiling at me with its familiar face, and the improbable sequence of events that have shaped our existence, a sense of wonder washing over me. This feeling goes beyond merry musings, grounded in a deep recognition of the universe's interconnected web. While the Moon isn't throwing popcorn from the cosmic bleachers, its mere existence attests to the awe-inspiring emergence performance, and we are all B-list celebrities.

Institutions

Sweaters and cathedrals. How can a symphony make you cry? Why do we see Jesus's face in toast? When we listen to music, what are we trying to do?

> *"A man's reach should exceed his grasp, Or what is a heaven for?"*
> *— Robert Browning*

If you have ever knit a sweater (or a really goofy scarf), you are part of a special club. A sweater is an everyday item and has simple emergent properties but is also an example of an innate need for something more.

According to Maslow's hierarchy of needs:
1. Physiological needs
2. Safety needs
3. Love and belongingness needs
4. Esteem needs
5. Self-actualization needs

A sweater could technically check off a number of these needs, but it could theoretically fulfil all of them. Why do human beings have such a deep desire for higher fulfilment? Churches, monarchies, and monuments. Religions, governments, and works of art. We have an innate longing to organize and build perfect, even holy, institutions. The need to belong to something bigger is very human. I would suggest our emergent mind reflects the emergent nature of this reality. If you have ever had the chance to visit a cathedral, you would know how art can inspire divine feelings in just about anyone.

"God created man in his own image…" — Genesis 1:27

The universe suddenly awoke. The universe and the information that defines it emerged. Everything in this universe follows this fractal blueprint of information. Information expanding into even more complex builds of itself spontaneously in one emergent event. The universe is emergent information, and the emergent process of life is embedded into the very fabric of our reality. This information permeates everything because everything is made from the expanding symmetry of this information. Emergence is fundamental to this reality and should be observed on all scales.

In this passage, I attempt to delve into the philosophical and metaphysical dimensions of the universe and its emergence, with a side order of poetry thrown in, to really complicate things unnecessarily.

At its core, this passage revolves around the concept of "Right Now" and aims to provide a glimpse into the interplay of time, fractal information, and collective consciousness. Through this exploration, we hope to illustrate how the emergent properties of our reality come to life.

We begin by contemplating how the universe suddenly "awoke." This awakening can imply its moment of creation, akin to the Big Bang, but it also resonates with the act of us being in the present moment or even waking up each morning.

We then ponder the universe and the information that defines it, emphasizing their emergence and preservation within the ever-present "Now." According to this perspective, the universe encompasses not only physical matter and energy but also the information that gives structure to its laws, properties, and history. Events like the Big Bang and past experiences now exist solely as informational artifacts, essential components for understanding our present reality.

In the context of this worldview, everything is viewed as information. The universe's information is likened to a fractal blueprint, where intricate patterns replicate at various scales. This implies that the universe's defining information continually evolves, leading to increasingly complex structures and phenomena. These developments are built upon a foundation of fractal information within the framework of time.

We proceed by discussing the universe's spontaneous emergence, highlighting its remarkable complexity arising without external influence, following its enigmatic course.

The emergent process of life (EPL) is introduced as a concept suggesting that life itself is an emergent property of the universe, rooted in the fundamental information and processes woven into the fabric of reality.

We stress the idea that fundamental information permeates everything, shaping all aspects of existence within the overarching concept of fractal reality.

In a rather poetic manner, we liken emergence to a fundamental force, akin to gravity, in the way it operates within the universe. This viewpoint contends that emergence and gravity are two complementary aspects of the same cosmic phenomenon, working harmoniously.

This passage wraps up by acknowledging that the emergent nature of the universe is expected to manifest itself at all levels of existence. This includes the tiniest subatomic particles and extends to the grandest cosmic structures, further emphasizing the recurring fractal pattern woven into the fabric of the universe.

I hope that this passage provides a philosophical perspective that beautifully weaves together information, emergence, and the universe's existence. It is meant to encourage us to reflect on the profound interconnections within the grand tapestry of reality. Furthermore, it invites us to question and explore the universal meaning present in our lives, emphasizing the power we hold to influence our world through even the simplest of gestures.

I'm not suggesting that this is the absolute truth, I'm just saying we should explore these ideas. It's essential to contemplate them because they aren't always crystal clear, and we shouldn't unquestioningly accept what we're told. So, we really have to dig deeper to find more profound meanings in our lives, explore our reality, and question our place in it.

People are constantly in search of meaning and, sometimes, life is simpler than we make it out to be. It's as straightforward as saying "good morning," being kind to others, and offering assistance. If everyone embraced these simple actions, it could have a remarkable impact. Sometimes, people underestimate how much control and influence they possess over everything around them and their place in it.

The Poetry of Emergence

In the vast tapestry of the universe, humanity emerges as a remarkable expression of intricate patterns and boundless possibilities. Our existence can be seen as a poetic dance between complexity and simplicity, where life's ordinary moments hold the potential for extraordinary transformation.

Consider a seed, small and unassuming, cradled in the earth's embrace. Within its fragile shell lies the promise of life, awaiting the perfect conditions to awaken its potential. This simple act of growth is akin to the emergence of something new from the old, a dance of renewal.

Our own lives follow a similar pattern. We start as unremarkable beings, yet within us, the potential for growth, change, and transformation resides. Like the seed, we require nurturing environments, experiences, and connections to flourish. Our emergence into unique individuals is a journey filled with wonder and surprise.

Think of a child taking those first tentative steps into the world; it's a moment of emergence, a transition from crawling to walking. Each stumble teaches balance, and each fall strengthens resolve until they emerge as confident explorers of their surroundings.

In our lives, we encounter countless moments of emergence. New friendships form, passions awaken, and insights unfold. These moments are like verses in the poetry of our existence, revealing the beauty of growth and change. They remind us that life is not a static destination but an ongoing journey of emergence.

Consider the bonds we form with others. These connections often start as simple moments or kind words, gradually deepening into sources of strength, love, and understanding. The emergence of profound connections between individuals is a testament to the power of human relationships and the beauty of emergence in our interpersonal lives.

Nature, too, is a master of emergence.

Seasons change, and with each passing day, we witness the transformation of landscapes. Spring emerges from winter's cold embrace, breathing life into barren trees. Flowers bloom, birds sing, and the world bursts with colour. This emergence of life, year after year, is a reminder of the cyclical nature of existence and the perpetual beauty of renewal.

In the grand tapestry of our lives, emergence is the thread that weaves through our experiences, connecting the dots and revealing the poetry of our existence. It's the process through which we grow, learn, and become. It's the magic that turns the ordinary into the extraordinary.

So, let us embrace the poetry of emergence in our lives. Let us recognize the beauty in our growth, the wonder in our connections, and the power of our transformations. Like a seed reaching for the Sun, we, too, have the potential to emerge from the soil of our past, reaching for the light of our future. In the unfolding story of our lives, emergence is the ink that writes the verses of our unique and beautiful journey.

Journey to Today

One Thursday morning, I found myself at a pivotal moment — facing my final exam, marking what I thought was the culmination of my educational journey. The day held an unforeseen twist — an acceptance letter from a lesser-known startup specializing in the production of commercial space satellites. This moment had been long-awaited, providing me with the opportunity to immerse myself in the fascinating world of manufacturing satellites for commercial space endeavours.

My fascination with space had been a lifelong companion, igniting my dreams and aspirations. This job offer was not just an opportunity, it was the realization of a lifelong dream. It marked my big break, an opportunity to immerse myself in the captivating world of outer space, and I had been chosen to be a part of this exciting journey.

However, there was a twist. The details of this thrilling chapter stipulated that I needed to report for duty the coming Monday in an unfamiliar and distant town. This added an extra layer of complexity to the situation. Undaunted, I packed my backpack and prepared for this new adventure.

One significant challenge was finding suitable accommodation for myself and my faithful canine companion. I needed a place that not only welcomed pets but also involved overcoming the financial hurdle of covering the first and last month's rent. To surmount this obstacle, I turned to my trusty credit card to bridge the financial gap.

On my very first day, I was introduced to "The Lab." It was a vast, pristine, big white box, an environment that would become my professional home for the next 28 years. This marked the beginning of my journey, and the path ahead was filled with potential, responsibilities, challenges, and opportunities for growth. It wasn't just a job, it felt like a calling. It was an adventure into the world of scientific exploration and the unknown.

In the midst of this new role, an unexpected twist of fate awaited. My new manager informed me that there wouldn't be any work for me in the next month. This made me nervous, and I could feel my financial house of cards beginning to shift.

My manager directed me to the HR department, where a unique opportunity awaited. I was to assist in a training project designed to improve management's interview skills, helping to select the best new employees. Over the following month, I found myself playing the role of a "guinea pig" for various management teams throughout the company, as they practiced interview techniques on me repeatedly.

Fate had dealt me a hand in middle management. It didn't happen all at once, but this is where it started, something I hadn't foreseen when I embarked on this journey. Sometimes we think we're looking straight ahead, but in truth, our lives twist and turn, leading us to the emergent events that brought us to this place we call today. It was a path filled with unexpected challenges and surprises, yet it also proved to be a journey of growth, learning, and, ultimately, a rewarding career in the world of science. It allowed me to form lasting friendships and discover what it meant to be a leader.

IS OUR EXISTENCE A MEANS
FOR THE UNIVERSE TO
COMPREHEND ITSELF?

Question 2

Is everything a function of consciousness and information?

In our quest for understanding, we ponder a fascinating question: is everything intricately connected through the harmonious interplay of consciousness and information?

Consciousness, that mysterious force that defines our very existence, stands at the crossroads where self-awareness meets the flow of information. It beckons us to explore its role in the vast expanse of the universe.

Alongside this enigmatic consciousness, we encounter fractal information — a constantly evolving blueprint that influences every facet of our reality.

Let's propose that Universal Consciousness and Fractal Information are the fundamental cornerstones of our existence. *Through their intricate dance, do they give birth to the tapestry of time, space, matter, and life?*

From this union springs the intricate web of time, the boundless dimensions of space, the substance of matter, and the intricate threads of life itself. This consciousness, a timeless essence we might call the Allspark, guides us through the cosmic narrative. Meanwhile, fractal information, an ever-evolving guide, shapes the contours of our existence in a perpetual story of discovery and growth.

> *"The human brain has 100 billion neurons, each neuron connected to 10,000 other neurons. Sitting on your shoulders is the most complicated object in the known universe."* — *Michio Kaku*

Information is not random.

Random is made, done, happening, or chosen without method or conscious decision (Oxford).

Information is facts provided or learned about something or someone (Oxford).

I prefer: Information is knowledge or data we gather and use to understand things and make decisions. The valuable content or facts help us learn, communicate, and make sense of the world. Information refers to the organization or arrangement of the elements within a system.

Random is not the exact opposite of information, but that point is academic. Information is not random.

> *"If a tree falls in a forest and no one is around to hear it, does it make a sound?"* — *Dr. George Berkeley*

Dr. George Berkeley, an Anglican Bishop and philosopher in the 1600s, pondered this thought-provoking question. Dr. Berkeley, known as the "father" of Immaterialism, believed that the answer to this question is affirmative. In his view, the tree produced a sound because God heard it. Immaterialism posits that material things have no inherent reality beyond being perceived as mental perceptions.

The inquiry surrounding the existence of a sound when an unobserved tree falls in the woods raises profound philosophical and physical inquiries.

Do things only exist if they are perceived? What insights can physics offer regarding the nature of reality?

These deep questions have remained unanswered for decades and, in some cases, centuries, fueling ongoing debates. Some may even question whether these inquiries can ever be definitively resolved.

As defined by Webster, immaterialism is a philosophical theory asserting that material entities hold no intrinsic reality except as they are perceived in the mind. This perspective challenges our conventional understanding of the physical world and offers an alternative lens to contemplate the nature of existence.

> *"I went to a bookstore and asked the saleswoman, 'Where's the self-help section?' She said if she told me, it would defeat the purpose."* — George Carlin

Consciousness is perhaps the biggest riddle in nature, and it's called "The hard problem of consciousness." Reduced to its simplest meaning, consciousness allows us to be aware of our surroundings and inner state.

Consciousness is the state of being awake and aware of one's surroundings (Oxford).

Thinking about consciousness is somewhat of a red herring. Consciousness is what we are currently observing right now. However, when we attempt to pinpoint the exact nature of this present moment, we struggle to define it and seek out mechanical or biomechanical explanations.

This pursuit has occupied philosophers and scientists for centuries as they engage in debates and search for the origins of consciousness. I claim that consciousness and information are universal... Yes, everywhere. Rocks, bodies of water, and empty boxes are composed of consciousness and information. What the inert lacks is the ability to observe.

SIDE QUEST

A red herring is like a tricky riddle in a mystery story. It's something that distracts you from the real clue or answer. Imagine you're trying to find a hidden treasure, and someone leaves a shiny, fake gem in your path. You might stop to pick it up, thinking it's the treasure, but it's just a distraction.

In real-life situations, we often encounter red herrings, especially when we're dealing with problems. Sometimes, the simplest solutions don't necessarily lead to the best answers, and we can end up going down the wrong path or hitting a dead end.

So, when you come across a red herring, remember to stay focused on what truly matters and not get sidetracked by distractions. It's a sneaky trick, but once you recognize it, you can keep your eye on the real prize or solution.

I believe consciousness permeates the universe and constitutes a fundamental aspect of it. I am not suggesting that everything is conscious in a literal sense. Rather, I mean that everything is composed of information, with consciousness serving as the inherent essence of this reality.

Imagine consciousness as the vast canvas of the universe, and fractal information as the brushstrokes that colour it. Each piece of information in the universe is like a contribution to the overall masterpiece. While individual brushstrokes aren't conscious, the canvas itself, embodying consciousness, forms the foundation of the entire artwork of reality.

Life mysteriously and spontaneously emerged 3.8 billion years ago. It began modestly but was constructed with an astounding fractal artifact and the emergent process of life (EPL). These life forms were simple, single-celled organisms that converted available energy into replicas. However, the remarkable fractal artifact was DNA (or RNA), propelling life on its transformative journey. The miraculous beginning of Earth's simplest life, powered by the incredibly complex engine of DNA, was like storing your grandmother's meatloaf recipes in a supercomputer or delivering pizzas with a Lamborghini. Still, regardless of how bizarre it may sound, this is where our story begins.

This odd beginning is an enigma, begging to question our origins and mystifying bright minds everywhere for time immemorial. The mystery of DNA was my first clue, and by looking at this question more closely, it propelled me on my journey of discovery, which I am truly passionate about sharing.

"The spiral in a snail's shell is the same mathematically as the spiral in the Milky Way galaxy, and it's also the same mathematically as the spirals in our DNA. It's the same ratio that you'll find in very basic music that transcends cultures all over the world." —— Joseph Gordon-Levitt

The first steps on this journey were primarily taken by simple organisms which looked like shapeless blobs and moved randomly and absorbed food particles they encountered along the way. They employed a simple strategy of speeding up in areas with low food availability and slowing down in areas where food was more plentiful.

Trichoplax adhaerens, or Trichoplax. Named after the Greek words for "hairy, sticky plate," Trichoplax belongs to one of the most ancient animal lineages on Earth, a phylum known as Placozoa that is more than 650 million years old. Trichoplax lacks nearly all the usual animal characteristics: it has no muscles, no stomach, and no neurons.

The next milestone on this journey was the active pursuit of food. Creatures like the Tiger flatworm had developed primitive sensory organs, and this peculiar-looking flatworm would follow scent trails to find its dinner. This type of flatworms, which often inhabit dark, watery environments shielded from direct light, lack complex eyes like ours.

However, many possess two lens-less primitive "eye pods" on their heads that can detect light intensity. These flatworms also have two ear-like lobes functioning as tactile and chemical sensors. Utilizing primitive perception abilities and a simple nervous system, they can employ a more complex strategy of moving towards food when hungry and seeking shelter in dark places when full.

Caution! Attention Required: I would have liked to describe this journey of observation using adorable kitten memes from the Internet. However, on this guided tour, I find it essential to highlight important clues. One of the initial clues is DNA. DNA is not just a clue, it's a question, an enigma. Life embarked on this journey of observation from a formless, shapeless, simple puddle, but this puddle was empowered by the enigma of DNA. This is an important clue, an outlier. I am not sure what it means but I am sure it is very important.

The next stage in the ladder of observation includes the evolution of spatial and temporal memory. The ability to perceive from a distance allows living organisms to remain aware of their surroundings. Life primarily focuses on obtaining energy and reproducing, which involves replicating its genetic information stored in DNA. The earliest steps toward awareness were likely taken by mobile creatures that moved toward food sources without a clear sense of their destination.

Across generations, life experimented with external sensors and developed larger, more complex neural networks. Vision contributes by adding context and depth to our world, enhancing our understanding of the space in which we and our sources of sustenance exist.

Consequently, the next improvement must occur internally.

To visualize food in its absence, an organism must create an inner representation of the world. This allows an animal to continue searching for food even when it is beyond its sensory range. By maintaining an inner representation of what is relevant in the world, it can remain focused on its food and its desire to obtain it.

The capacity to remember emerges, enabling animals to be briefly distracted from their pursuit but quickly resume their path afterward. A related phenomenon known as object permanence arises, signifying our awareness that things continue to exist even when out of sight. This cognitive skill is present in animals (cute kittens), birds, insects, and some invertebrates. The ability to remember something in its absence suggests at least a basic sense of time. A sense of time marks a significant stride on the journey towards awareness and allows a self to look ahead from the present moment and anticipate the future.

Delayed gratification leads to an understanding of increasingly complex perspectives. The ability to comprehend the wants and needs of other organisms is vital for higher levels of awareness. Communication takes this ability to understand others and represent the absence to an entirely new level, eventually giving rise to language, music, and enhanced expressive communication.

So, back to the red herring... consciousness is not something we are doing right now, it is something we are experiencing. Our physical self allows us to perceive consciousness.

Our physical self allows us to perceive consciousness (This is important!)

Science says we should always go with the simplest explanation. Well, it doesn't get any simpler than this. Why are we conscious? Because the universe is." — *Unknown*

ODD THOUGHTS

In the intricate tapestry of existence, I hold the belief that consciousness is not limited to humanity alone. It's a thread that weaves through the fabric of life, connecting all living beings.

Whether it's the towering presence of ancient trees, the intricate world of buzzing insects, the keen awareness of jumping spiders, or the unwavering loyalty of our canine companions, there's a shared essence of consciousness that unites us all. It's a reminder that we are part of a greater whole, coexisting and coevolving with the diverse array of life on this planet. This recognition of consciousness in every corner of nature inspires a profound sense of interconnectedness and responsibility to nurture and protect the web of life that surrounds us.

So what is next in our journey? ... Intelligence.

Intelligence is the next milestone in our journey. As life evolved, organisms gradually developed the ability to process information, learn from experiences, and adapt their behaviour accordingly. Intelligence can be understood as the capacity to perceive, understand, and solve problems, and it is not limited to human beings alone.

The emergence of intelligence can be seen in various forms throughout the animal kingdom. From the problem-solving skills of primates to the navigational abilities of migratory birds, intelligence takes on different shapes and sizes. It is not solely dependent on brain size or complexity but instead on the efficiency of information processing and the ability to respond flexibly to the environment.

As we delve deeper into the realm of intelligence, we encounter fascinating examples. Cephalopods, such as octopuses and squids, exhibit remarkable problem-solving skills and a high degree of behavioural flexibility. These creatures have complex nervous systems and demonstrate a capacity for learning and memory.

On the journey towards intelligence, social interactions play a crucial role. Cooperation, communication, and the exchange of information between individuals contribute to the development of collective intelligence. For instance, social insects like ants and bees exhibit sophisticated division of labour and coordinated behaviours that enhance survival as a group.

Human intelligence stands out as a remarkable culmination of this journey. Our ability to reason, imagine, create, and communicate through language has propelled us to incredible achievements. We have harnessed the power of technology, created complex societies, and expanded our understanding of the world.

As we embark on our ongoing journey, the exploration of intelligence holds the potential to lead us to new frontiers. It allows us to further unravel the mysteries of consciousness, uncover the principles underpinning collective intelligence, and push the boundaries of what is possible. I believe our quest for intelligence is not solely driven by innate curiosity and the desire to understand ourselves and the world around us but also reflects the very nature of our reality.

Therefore, as we contemplate what lies ahead, I believe it is crucial to recognize that our journey toward intelligence is deeply intertwined with the intricate tapestry of reality itself. From the humble beginnings of single-celled organisms to the complexity of human consciousness, each step has paved the way for the next. As we continue to venture forward, the exploration of intelligence promises new horizons of discovery and understanding, ultimately allowing us to grasp the profound interconnectedness between consciousness, shared fractal information, and the purposeful nature of our universe.

So, why is intelligence important?

The anthropocentric paradox challenges the notion of intelligence's importance. Some argue that intelligence may not hold significance, questioning whether the vast eternal universe and most lifeforms lack intelligence. However, this argument arises from a particular perspective: anthropocentrism. The belief in human centrism, which dates back to the beginning of humanity, asserts that humans are separate from, and superior to nature. It propagates the idea that other entities, such as animals, plants, and minerals, exist solely as resources for human use.

This perspective is deeply ingrained in many aspects of our modern world, including cultures, political structures, businesses, and religions, with much of our verified information aligning with this viewpoint.

But can we challenge the anthropocentric view?
Is it possible to see the world from a different lens?
Is there something more beyond this perspective?
This is where the paradox lies. What is our argument against this view?

Our faith in a higher power?
The contemplation of extraterrestrial life?
Our very human philosophical disdain for selfishness?
Is it possible that our universe and human life possess a deeper purpose or meaning, rather than being a random occurrence without a centre or significance?

Merely dismissing the anthropocentric view as an illusion or mistake seems inadequate. There must be more to our existence. This paradox prompts us to question our worldviews, political ideologies, religious beliefs, and even the ever-evolving landscape of science.

It reminds us that we perceive ourselves as special, yet our actions often fall short of that nobility. It begs the question: *why do we, if we are truly noble, behave poorly?*

The answer is... choice.

I believe intelligence becomes important because it grants us the power of choice. It provides us with the ability to make decisions, to reflect on our actions, and to strive for a higher standard. Intelligence allows us to transcend our base instincts, to question, learn, and grow. It enables us to shape the world around us and contribute to its betterment.

In essence, intelligence is significant because it presents us with the opportunity to rise above our inherent flaws and make choices that align with a broader, more compassionate understanding of ourselves, others, and the world we inhabit. It is through the exercise of our intelligence that we can work towards a more enlightened, empathetic, and responsible existence.

The first few times I glimpsed the profound mysteries of consciousness, birth, and death, my emotions swirled in a tapestry of awe and wonder.

The initial encounter with consciousness came with the arrival of my son into this world. It was an experience that defied simple words. When my daughter was born, I was graced with that profound sensation once more. In those precious moments, I realized that consciousness transcends mere existence. It's more than just drawing breath, it's an ethereal, intangible essence that infuses life with meaning. It's about sharing the tapestry of this reality, a connection that goes beyond the physical.

Then came the poignant moment when my beloved grandfather departed from this earthly realm. His departure was swift and sudden, leaving us all in shock. It was as if a cherished flame had been extinguished in the blink of an eye, leaving behind a profound void. In that instance, I realized that my understanding of myself and the universe was far from complete. My journey had only just begun.

As the years rolled on, it wasn't until I reached my forties that the pieces of the puzzle began to align. I started to discern the bigger picture, to grasp the intricate web of life, consciousness, and the profound significance of each fleeting moment. These experiences painted my existence with hues of wisdom and gratitude, leaving me humbled by the vast mysteries that still await discovery.

The Anthropic Principle
The Anthropic Principle is the principle that the universe is constrained by the necessity to allow sentient life.

The Anthropic Principle, influenced by Renatus Cartesius, suggests that the way the universe is organized allows conscious beings to exist, akin to Cartesius' idea: 'I think, therefore I am.

To observe the universe, a conscious observer is necessary.

Although it may appear straightforward, a particular type of universe is required to host a conscious observer — one capable of developing and sustaining sentient life for a sufficient duration to allow the emergence of an advanced civilization capable of observation. Examining our own history reveals that this is not as simple as it sounds. Earth has supported life for approximately 3.8 billion years, yet an advanced civilization with significant meaning has only existed for less than 25,000 years.[6]

Proponents of the anthropic principle argue that it offers an explanation for why our universe possesses the age and fundamental physical constants essential for the existence of conscious life. Without these factors, we would not be here to make any observations. Anthropocentric reasoning often addresses the concept of fine-tuning in the universe and can be

---

[6] It's essential to note that the study of Earth's history and the search for potential advanced civilizations in deep time are subjects of ongoing scientific research and exploration. New discoveries and evidence may emerge in the future that could shed more light on this topic.

categorized into the "weak" and the "strong," based on their cosmological claims.

The weak anthropic principle (WAP) suggests that the perceived fine-tuning of the universe arises from selection bias, particularly survivorship bias. Some arguments incorporate the notion of a multiverse to expand the statistical population of available universes. Nevertheless, a single vast universe generally suffices for most forms of the WAP that do not specifically address fine-tuning.

SIDE QUEST
In a multiverse, think of it like rolling a cosmic dice over and over again. With each roll, you create a new universe with its own unique events and possibilities. So, anything can happen if you roll the dice often enough in the multiverse, just like different Marvel storylines can unfold in their own universes.

The strong anthropic principle (SAP) posits that our universe is meticulously fine-tuned to support conscious life. It proposes that the fundamental laws and conditions of the cosmos are precisely orchestrated to allow the emergence of intelligent beings who can contemplate their existence. Advocates argue that even slight alterations to the universe's parameters would render life impossible. They contend that the universe possesses a drive to evolve toward states capable of sustaining conscious observers.

The Conscious Anthropic Principle (CAP) is undeniably a variant of the strong anthropic principle, as it postulates that the universe is not a random assembly of particles and energy, but rather a purposeful creation that deliberately fosters the development of intelligent life. According to this theory, consciousness interacts with shared fractal information, shaping the universe and generating the finely tuned conditions we observe. It proposes that the universe is not a mere coincidence but a deliberate creation driven by conscious intentionality and constructed using shared patterns of fractal information.

Anthropic "coincidences" — Anthropic Bias by Nick Bostrom

Nick Bostrom's book *Anthropic Bias* explores the idea of anthropic "coincidences" in our understanding of the universe. These coincidences are like remarkable and specific conditions that make life possible. Imagine these conditions as being extremely precise, like hitting a tiny target on a dartboard. If these conditions were even a little different, life, especially intelligent life, couldn't exist.

One example comes from Robert Dicke's research in 1961. He noticed that the age of the universe seems just right for life to exist. If the universe were much younger, there wouldn't be enough time for the right elements to form planets. And if it were much older, stars would have burned out, making life impossible. This special age of the universe is a bit like finding a "Goldilocks zone."

Dicke also figured out that the amount of matter in the universe must be just perfect to prevent a Big Crunch. This is referred to as the "Dicke coincidences."

Other things, like the strength of fundamental forces in the universe, need to be just right for life. If they were even a bit different, life as we know it couldn't happen. These values are incredibly precise, like tuning a musical instrument to get the right sound.

In a remarkable display of scientific insight, Fred Hoyle predicted the existence of an excited state of the carbon-12 nucleus at a specific energy level. This insight led to an understanding of how helium-4 nuclei combine to form carbon inside stars, a crucial process for life and a remarkable illustration of the anthropic principle in action.

> *"I never made one of my discoveries through rational thinking."*
> — *Albert Einstein*

This serves as yet another remarkable clue. When we reverse-engineer scientific observations with an anthropic point of view, we can deduce scientific facts, much like a compass guiding us in the right direction.

This is akin to observing a baker's oven and deducing that its purpose is to create delicious bread. Just as the oven's function becomes evident when we see the bread it produces, understanding the Sun's role in synthesizing elements essential to life like carbon helps us comprehend its significance in the grand scheme of life's existence.

So, these anthropic coincidences show that our universe seems finely tuned for life. If any of these precise conditions were different, life - especially intelligent life - wouldn't be possible. It's like the universe has a set of very delicate dials, and if you turn them even a tiny bit, the whole picture changes, and we wouldn't be here to talk about it.

Let me summarize what we think we know:

- The Earth was formed 4.5 billion years ago.
- Fossils from 3.8 billion years ago reveal primitive life. DNA or RNA.
- DNA is an extraordinarily complex molecule and contains all the necessary ingredients for the emergent process of life (EPL).
- 3.8 billion years ago the Earth cooled to a point where liquid water could exist.
- Life has emergent properties.

- Life was observed on Earth as soon as it was possible. Instantaneously (give or take a million years — which is actually quite short but much longer than total human history).
- We only have one tree of life. All life on Earth is descended from this one occurrence, and in the last 3.8 billion years, it has never happened again.
- Successful life has more offspring.
- The emergent process of life is moving towards observation.
- Intelligence allows life to have a richer and fuller experience of consciousness.
- Human beings seem to have free will or choice.
- The universe we observe seems perfect for us, which implies an anthropic bias.

The observable universe appears remarkably suited to our existence. This bias stems from our physical attributes, which enable us to perceive consciousness. Abilities such as observation, intellectual reasoning, and the capacity to retain spatial and temporal memories are essential for the meaningful experience of consciousness.

5 FORCES

This is only our first peek down the rabbit hole, and we are only getting started. Science, math, and physics are wonderful inventions of human ingenuity, but we must always be mindful of our limitations and realize our own shortcomings. Dogma, comfort, and our ever-present biases will bind us and slow us down.

I ask you to be skeptical of what you think you know and look for the truth. Most of your beliefs are not even yours. Knowing that, you can more easily let them go.

In our journey through the intricate tapestry of existence, we encounter a profound confluence of religion, philosophy, and science. Once, these pillars stood in harmonious unity, guiding humanity towards a deeper understanding of life's mysteries.

However, as we navigate the intricate realm of religion today, we find ourselves facing a significant challenge. Religion, while offering invaluable insights to individuals and communities, sometimes casts a shadow on the path of inquiry. Yet, it's crucial to recognize that religion has always been a beacon of love and understanding, illuminating the moral landscape. It has played an indispensable role in enriching our comprehension of existence and fostering empathy and compassion among us.

I hold the belief that religion, in some profound way, appears to be intrinsic to our human essence. This essence is so deeply ingrained that even some atheists argue their points with a fervour akin to a spiritual crusade. It's likely that religion had its origins in the early human mind's cognitive ability to be able to formulate ideas and extract answers from the enigmatic cosmos. This innate tendency to extrapolate information, which we witness in religious beliefs, could very well be an evolutionary trait. While acknowledging the challenges it may present, it's not beyond our capacity to reconcile this aspect of our nature with the ever-evolving landscape of human knowledge and understanding.

We live in an absurd world.... Don't get me wrong, I really like this world, but the love and beauty of this place do little to explain it.

How does the whimsical labyrinth of our reality reflect the logical framework of information and consciousness, and what can we deduce about our place within it?

It was once thought that our reality has four fundamental forces (five if you include love): strong, weak, electromagnetic, and gravitational.

Welcome! I am Odd, your tour guide. Let's begin with some foundational knowledge that will hopefully empower you to ask the big questions. I invite you to expand your mind and absorb all the information you can. Below is a concise overview of the various components provided in your assembly, serving as a new perspective for this reality and guiding you to understand the rules of our existence.

I've broken down the complex steps into quick one or two-sentence summaries, followed by more detailed descriptions for those who are up for the challenge or who find a particular subject intriguing. This format allows you to delve as deep as you desire, making it suitable for both quick reference and in-depth exploration. It can also be used as a reference guide with easy-to-understand descriptions on some of the universe's complicated puzzles.

My goal is to suggest everything is informational in nature and to hint that the storing of information in the present moment has profound implications.

I've handled most of the heavy lifting, so let's get started!

The Strong Force
The strong force is a fundamental force that holds atoms together.

The strong nuclear force is a bit like a slipknot. Imagine you have a rope with a slipknot tied in it. No matter how hard you pull on the ends of the rope, the slipknot keeps the loop together. Similarly, the strong nuclear force keeps the particles in an atom's nucleus bound together, even when they should be pushing apart due to their electric charges.

The strong force operates within the nucleus over a short range, exchanging particles called quarks and gluons between nucleons to transmit the force. Quarks are elementary particles, while gluons act as carriers of the strong force, similar to how photons (carriers of light) transmit the electromagnetic force.

An intriguing property of the strong force is its "asymptotic freedom," becoming stronger as nucleons get closer to each other (slipknot). The force tightly binds them together. However, attempting to pull nucleons too far apart results in an even stronger force, preventing their separation. This behaviour is known as "confinement."

The strong force follows the simple fractal blueprint of our universe, governing the construction of matter. It is essential for the stability and structure of matter, enabling protons and neutrons to form atomic nuclei, which serve as the fundamental building blocks of all elements. Without the strong force, the world as we know it would not exist.

The Weak Force
The weak force is one of the fundamental forces in nature. It is responsible for how atoms and subatomic particles behave.

Imagine fundamental forces are like personalities: The strong nuclear force would be the tough and sturdy bodyguard, always keeping things tightly together. Electromagnetism would be the outgoing and attractive socialite, drawing things together with a spark. Gravity would be the friendly and dependable neighbour, keeping everything in its place.

Now, the weak nuclear force? It would be the shy and subtle scientist, quietly working in the background, making small but incredibly important discoveries. It's not one to grab the spotlight, but it plays a crucial role in the delicate balance of particles and forces in the universe.

Think of the weak force as a delicate hand that mainly interacts with tiny building blocks of matter, like quarks and leptons. This force comes into play during processes such as radioactive decay, where particles can transform into others, similar to how a chameleon changes its colour to blend into its surroundings.

In this subatomic world, the weak force has its own messengers, W and Z bosons, which act like couriers mediating interactions between particles involved in weak processes. Unlike other forces, such as electromagnetism, which can reach across long distances, the weak force operates only at extremely short distances, akin to how a whisper can only be heard up close.

Understanding the weak force is like deciphering a hidden code in the world of particles. It helps us explain peculiar types of radioactive transformations, much like uncovering the secrets of a cryptic puzzle, and provides a window into the behaviour of matter at its most fundamental level, similar to peering through a microscope at the tiniest details of a painting.

The Electromagnetic Force

One of the most observable of the fundamental forces in nature. It is responsible for the interactions between electrically charged particles, such as electrons and protons.

Imagine the electromagnetic force as a cosmic storyteller with a vast toolbox. In this toolbox, it has many magical tools, each serving a unique purpose.

1. Light: One of its most enchanting tools is light, like a beam of knowledge that brings stories from the stars.
2. Electricity: Another essential tool is electricity, like a swift messenger that powers our devices and delivers modern tales.

3. Magnetism: It also wields magnetism, like a captivating force that attracts and repels, guiding us like a compass.

4. Radio Waves: Picture radio waves as its melodious voice, broadcasting stories through the airwaves for us to hear.

5. Microwaves: Microwaves act as its gentle warming touch, cooking up stories in the kitchen, making popcorn tales come to life.

6. X-rays: X-rays are like its curious eye, peering into hidden narratives, revealing the secrets beneath the surface.

7. Infrared: Infrared rays serve as its warm embrace, revealing the heat of stories in the dark, like a cosmic campfire.

So, the electromagnetic force weaves its stories, illuminating our understanding of the universe, and serves as a power source for things like lights, cell phones, and electric motors. It's what makes these everyday experiences possible.

The electromagnetic force manifests itself in two distinct but interconnected aspects: electric fields and magnetic fields. Electric fields arise from charged particles and exert forces on other charged particles. For example, positively charged particles attract negatively charged particles, while like charges repel each other.

On the other hand, magnetic fields are produced by moving electric charges, such as currents in wires. These magnetic fields can exert forces on other moving charges, either attracting or repelling them depending on their relative orientations.

Electromagnetic force plays a central role in a wide range of phenomena and applications. It governs the interactions between charged particles, giving rise to electricity, magnetism, and electromagnetic waves. It is the force behind the functioning of electrical circuits, the generation of light, and the operation of devices such as motors, generators, and transformers.

In addition to its practical applications, the electromagnetic force is responsible for the structure and stability of matter. It determines the arrangement of electrons around atomic nuclei, giving rise to the chemical properties of elements and the formation of molecules.

The behaviour of the electromagnetic force is mathematically described by Maxwell's equations, which provide a comprehensive understanding of electric and magnetic fields and their interplay. Overall, the electromagnetic force is essential for the functioning of our technological advancements and plays a fundamental role in shaping the physical world as we know it.

BLACKSMITH'S TALE

I love taking courses, and I love the opportunity to learn. One such opportunity was taking a series of blacksmithing courses. I love the material, the heat, the metal, and the way you can shape the steel. The power and might, along with the utility of steel, have always fascinated me.

In this course, I learned many things, along with earning many blisters. I also noted that blacksmiths use a magnet to determine if you are at the proper temperature to be quenched — the critical transformation temperature or Curie point.

The question was asked: What is magnetism, and why does it go away when it gets hot? I love questions; they fascinate me, and I also knew the answer.

During lunchtime, we all took a break together, and I asked the group if they would like to know what magnetism was. Everyone was engaged and genuinely wanted to know the answer, including many of the older blacksmiths who were thirsty for this information.

I told them I would need an assistant to demonstrate what magnetism was. A young and brave future blacksmith named Fish volunteered for this assignment.

I began by sharing a simple truth: nature dances to the rhythm of four fundamental forces — five, if you count love (with a wink). The strong force binds atoms, the weak force orchestrates chemical unions, the electromagnetic force, which we're more familiar with, commands radios, TVs, light, and electricity, and last is gravity. My focus, however, was on the electromagnetic force.

I explained that for a material to be magnetic, it needs three things: shape, movement, and power.

Regarding shape, only a few atoms in the periodic table are small enough to be affected by a half-filled electron shell. In fact, only three of these small atoms are built in a lopsided manner that causes them to act like tiny magnets.

I asked Fish to make a fist with one hand. This fist represents the nucleus of the atom, the positive part of the atom. Then I asked Fish to cover the top of the fist with the other hand, creating a lopsided ball. This half-circle represents the half-filled electron shell that makes up an iron atom. The half-circle of electrons forms the negative part of the atom. Does this lopsided shape already look like a magnet?

Next is movement. When we heat up the iron and cool it down quickly, or when we use tools like hammers or place the iron near a magnetic field, the tiny magnets inside the iron move around and change their arrangement. This rearrangement happens in a special structure called a "crystal lattice." Iron has crystal structures or matrixes and can move from one configuration to another when heated. The way these tiny particles move from one structure to another is like changing neighbourhoods, and it's what makes the iron more or less magnetic.

Alright, Fish, it's your turn! Make two fists and put them together. Look at how Fish's fingers are all jumbled up and mixed, that's how iron atoms usually look. Now, watch as we bring the iron close to a magnet.

Fish, open your hands so that all your fingers are pointing up. Now you can see that all the fingers, which are like atoms, are lined up. This alignment of iron atoms is what gives iron its special magnetic properties.

Now, let's talk about power. Fish, I want you to jump straight up into the air. With a big smile, Fish jumped as if trying to touch the ceiling. What Fish just did was push against the Earth, and Fish was strong enough to beat the whole planet! (Big cheers!)

The electromagnetic force is way stronger than the force of gravity. In fact, the electromagnetic force is about 10^{36} times more powerful than gravity. Because of this big difference in strength, magnets and our buddy Fish seem really strong.

We finished lunch and went back to the forge. I think my hammer felt a little lighter.

GRAVITY

Gravity is what mass does.

All physical things have mass, and it's this property that distorts this place we call home. It's a fundamental aspect of the universe, with a tendency to pull everything together. I believe there's an anthropic reason for gravity's needy personality. Without gravity, we wouldn't have planets, stars, or anything at all. Gravity is what keeps us together. This aspect of our reality reflects how all matter sticks together, displaying a personality that mirrors our reality's fundamental desire to live. Gravity is the personality of mass, and like a personality, it naturally pulls other personalities together and can be described as a tendency of nature.

Gravity is the phenomenon that we perceive as an attractive force, which keeps our feet on the ground, the moon in the sky, and locks our planet in orbit around our star, the Sun. Gravity is not the warping of space-time. While it may be convenient to illustrate objects falling into gravity wells, it is simply not the case. Time is not a dimension, and gravity is not a force.

So, what is gravity?

In a serene garden, an apple tree stood near Sir Isaac Newton's study window. One day, as the story goes, a ripe apple fell from the tree and landed near his window. This simple event, whether embellished or not, inspired Newton to contemplate the force that governs such falls. From this reflection, he developed his groundbreaking theory of gravity, forever altering our understanding of the world.

For millennia, we have studied the effects of gravity, perceiving it as a force. However, it is actually the by-product of mass interacting with reality. I would like to suggest that our reality could be seen as composed of numerous tiny fractal pockets that follow simple rules and contain lists of information, operating without force carriers. These pockets would work in concert, holding all the necessary information for the present moment. The information within these pockets is fractal in nature, adhering to a universal blueprint that determines potential outcomes.

Energy and matter exist as information within these pockets, encompassing various elements, of which mass is merely one. The format of these elements is defined within the fractal pocket. Fractal pockets share information among each other and abide by straightforward rules without the need for force carriers. Therefore, the force we perceive as gravity is simply information being applied.

The force we perceive as gravity plays a crucial role in enabling nuclear fusion to occur in stars, including our very own star, the Sun. Gravity is responsible for compressing the core of a star, increasing its temperature and pressure to the necessary levels for nuclear fusion to take place.

In the core of a star, gravitational pressures pull inward, causing the hydrogen gas to be squeezed together. As the gas becomes denser, the temperature and pressure rise. At a certain threshold, known as the ignition temperature, the pressure of gravity becomes strong enough to overcome the repulsive force of the weak force, allowing nuclear fusion to occur.

During nuclear fusion, hydrogen nuclei (protons) collide and fuse together to form helium. This process releases an enormous amount of energy in the form of light and heat. The energy generated through fusion counteracts the force of gravity, maintaining the balance within the star.

So, in essence, gravity is essential for initiating and sustaining the conditions necessary for nuclear fusion in stars and allowing their life-giving light. Gravity overcomes the repulsive force between protons, enabling them to come close enough for the weak force to facilitate fusion reactions.

LOVE

Knowledge is the foundation of our existence. Love is its purpose.

Love is a complex and multifaceted emotion that has captured the attention and imagination of humanity throughout history. It is a fundamental aspect of the human experience, intricately intertwined with our reality.

At its core, love represents a deep and profound emotional attachment and connection. It transcends boundaries and unites individuals, fostering a sense of belonging and intimacy. Love can manifest in various forms, including romantic love, familial love, platonic love, and even a broader love for humanity and the natural world. It is a force that binds us together and gives meaning to our lives.

Love is characterized by genuine care, compassion, and empathy towards others. It goes beyond self-interest and superficiality, seeking to understand, support, and cherish the well-being of those we love. Love often requires acts of selflessness, sacrifice, and a willingness to prioritize the needs and happiness of others above our own.

Furthermore, love is closely intertwined with acceptance and forgiveness. It acknowledges and embraces the imperfections, flaws, and differences in others, choosing to love unconditionally. Love promotes empathy, understanding, and the ability to see the world from another person's perspective, fostering harmonious connections and deepening our understanding of consciousness.

The power of love extends beyond mere emotion. It has a transformative effect, inspiring personal growth, healing, and the development of meaningful connections. Love nurtures trust, intimacy, and vulnerability, creating an environment where individuals can authentically express themselves without fear of judgement. Through love, we have the opportunity to evolve and expand our consciousness, contributing to the collective growth and well-being of humanity.

Love is not confined to the realm of emotions alone, it is a force that drives action. Kindness, affection, support, and acts of service are tangible expressions of love in our interactions with others. Love finds its voice through gestures, words, and deeds that communicate care, appreciation, and devotion.

It is important to acknowledge that love is not immune to challenges and complexities. Relationships and love require effort, effective communication, and a commitment to understanding and resolving conflicts. Love encompasses moments of growth, change, and occasional heartache. However, true love perseveres, seeking reconciliation and maintaining the connections and bonds between individuals.

Love is an emergent force and the high point of human existence. Love is a profound and transformative force that allows us to be more than we are. Love unites individuals, fosters connection, and promotes personal and collective growth. It is expressed through both feelings and actions, enriching our lives and contributing to the ongoing development of our shared reality.

THE SMALL: THE ENIGMA OF QUANTUM MECHANICS

Our current models, limited in scope, adhere to a simplistic paradigm that necessitates the division of our understanding into three distinct systems: the realm of the small, the domain of the living, and the expanse of the large.

THE SMALL

The small is the quantum world and the realm of atoms and fields. We naively accept that things at this scale don't have to behave the same way as objects on the larger scales and the weirdness is truly odd.

Think of quantum as a pocket-sized world where particles, called quanta, behave in unique ways. These quanta come in bundles of specific sizes or levels, like coins in different denominations.

In this tiny pocket, these particles can exist in multiple places at once and become entangled, just like magical coins that can teleport or affect each other's state, no matter how far apart they are. Quantum physics explores the rules of this intriguing pocket-sized realm.

The Atom
An atom can be thought of as the fundamental building block of all things.

An atom can be compared to an individual Lego block in a construction set. It has a central nucleus, similar to the central piece of a Lego structure, housing protons and neutrons. Electrons, with their negative charge, encircle the nucleus, much like Lego pieces connecting to the central block.

Protons
Protons are like the unique features or designs on individual Lego blocks. Just as these features distinguish one Lego piece from another, the number of protons in the nucleus defines the element's identity. They give atoms their distinct characteristics, just as unique designs make each Lego piece stand out.

Neutrons

Neutrons serve as the stabilizing elements that connect Lego pieces together in a structure. Just as interlocking Lego blocks require a certain balance to maintain the structure's integrity, neutrons in the nucleus help stabilize it by interacting with protons. The total number of protons and neutrons in the nucleus determines the atom's mass.

Electrons

Electrons are subatomic particles with a negative electric charge that orbit the nucleus of an atom. They play a crucial role in chemical reactions and electrical conductivity, as their arrangement in electron shells determines an atom's chemical properties and ability to form bonds.

Remember in science class when we were told electrons circle their atoms in orbits? In truth, we believe electrons do not orbit nuclei; they are arranged in pockets of probability in alternating subshells in a corrugated dance. The arrangement is closer to a Dewey Decimal system of books in a library than an orbit. We can look up the place the electron is likely to be, but we need to go and find the electron ourselves to be sure.

If you were to take any tiny piece of matter in our known universe and break it up into smaller and smaller constituents, you'd eventually reach a stage where what you were left with was indivisible. Everything on Earth is composed of atoms, which can further be divided into protons, neutrons, and electrons. While protons and neutrons can still be divided further, electrons cannot.

The concept of the size of an electron is also a bit different from what we typically understand as the size of a physical object. Electrons are elementary particles, which means they are considered to be point-like particles without any internal structure and are informational in nature. As far as our current understanding goes, electrons are considered to be fundamental particles with no size in the classical sense.

ENERGY FIELDS

In simple terms, "fields" are invisible areas of influence that exist all around us. They are like invisible force fields that can affect objects or particles within their reach.

Imagine energy fields as something like a magnetic or gravitational bubble surrounding an object.

Different types of fields exist, such as magnetic fields, which can attract or repel magnets, or gravitational fields, which cause objects to fall toward the Earth. Electric fields, for instance, can make your hair stand on end when you rub a balloon on your head.

These fields are crucial in understanding how forces and interactions work in the natural world. They help explain why objects move, how magnets attract or repel, and even how your devices can connect to the internet through wireless networks. Though we can't see them, fields play a significant role in shaping our physical reality.

Are they imaginary?

Fields are not imaginary in the sense that they don't exist, while we cannot directly see or touch fields, their effects are very real and measurable. Fields are a fundamental concept in physics and are in my opinion simply a manifestation of information.

MOMENTUM

Momentum is like the oomph factor of an object in motion — it embodies the intensity of its push or pull. This dynamic property is the result of both the object's mass, signifying how much "stuff" it contains, and its velocity, indicating how fast it's moving. Momentum can be thought of as information of movement.

Momentum is a core physics concept explaining an object's motion and its resistance to change. It relies on an object's mass and speed and is vital for understanding motion and interactions.

In the case of light, momentum is relevant to photons, which are massless particles of light possessing energy and speed. Photons can have a type of momentum due to their incredible speed. This photon momentum leads to various effects, including exerting forces such as light pressure on surfaces in the vacuum of space. This phenomenon contributes to various interesting occurrences, such as radiation pressure and the Compton effect in this unique environment.

SIDE QUEST: QUANTUM ESPRESSO
When ordering you can only order in whole numbers. No half-shots, in-betweens, long or short allowed. Welcome to the quantum cafe!

Now, let's simplify the concept of quantization and pixelization:

Quantization is like climbing a staircase where each step has a specific height. You can't stand between steps, you're either on one or the next. In the quantum world, properties like energy or momentum work the same way — they can only have specific, fixed values, with no in-betweens. This helps explain how tiny particles like electrons behave, and it's quite different from what we experience in our everyday lives.

Pixelization, on the other hand, is what you see on a digital screen, like your phone or TV. Every image is made up of countless tiny units, or pixels. It's as if the universe is a giant puzzle, and each piece (or pixel) holds a piece of information. This idea raises questions about the nature of reality and how everything fits together.

So, while quantization takes us into the mysterious world of quantum mechanics, where properties are fixed in specific values, pixelization makes us think of the universe as a vast digital image with tiny, interconnected pieces. Both concepts challenge our understanding of reality and lead us on journeys into the unknown, inspiring curiosity and imagination.

Could quantized particle properties be linked to a deeper cosmic structure, a coded fractal blueprint?

PHOTONS ARE THE CARRIERS OF LIGHT

Photons are truly extraordinary particles. These minute and indivisible units of electromagnetic waves possess distinct attributes — zero mass, no electric charge, yet capable of carrying discrete and indivisible packets of energy. Visualize a photon as a tiny bundle of information, a manifestation of the quantum nature where energy and matter are quantized. What sets photons apart is their intriguing wave-particle duality, a concept that continually challenges and captivates scientists.

Moving to the atomic and molecular realm, where discrete energy levels resemble steps on a ladder, each step corresponding to a specific, quantized amount of energy. Photons, particles of light, also carry energy. When a photon interacts with an atom or molecule, it engages in one of two fundamental processes: absorption or emission.

In the absorption process, an atom or molecule takes in a photon, transferring its energy. This causes electrons within the atom or molecule to ascend to higher energy levels, akin to climbing a ladder. This absorption changes the state of the atom or molecule, rendering it in an excited state.

Conversely, when an excited atom or molecule releases a photon, it signals an electron moving from a higher energy level back down to a lower one. This transition releases excess energy as a photon, emitted as light.

ODD FACT

LED lights and solar panels use this ladder trick to emit or absorb light (photons).

The specific energy levels of atoms and molecules dictate the colours or wavelengths of light they can absorb or emit. Each has a unique set of energy levels, leading to varied absorption and emission patterns (LEDs come in different colours…).

The absorption or release of photons involves the transfer of energy from light particles to atoms or molecules, resulting in changes to their energy levels and the emission or absorption of light.

Think of photons as tiny particles of light or pockets of information. They travel as waves through space, created by electric and magnetic fields that wiggle side to side, akin to a waving flag. These fields self-propagate, enabling light to travel even through empty space.

What makes photons fascinating is their dual nature — they act as concentrated bits, similar to particles, with momentum and specific properties. Simultaneously, they behave like waves, exhibiting specific frequencies and wavelengths, similar to ocean waves. Photons are a unique blend of particle and wave behaviour, adding to their intrigue in the realms of light and energy.

The wave-particle duality of photons manifests depending on context and observation, deepening the enigma surrounding these remarkable particles. Think of photons as actors on a stage. When spotlighted (observed), they play their role as particles, with defined properties. But in the dark (unobserved), they become like waves, moving and spreading out, creating a mysterious, hidden performance.

The Double-Slit Experiment

This is a fundamental physics experiment that involves firing particles or light through two narrow slits. It demonstrates that particles exhibit both wave-like and particle-like behaviour, depending on whether they are observed or not, challenging our understanding of the duality of matter and the role of observation in quantum mechanics.

The double-slit experiment was first performed by Thomas Young in 1801. This now famous demonstration that light and matter can behave both as waves and particles. The idea behind the double-slit experiment is to send particles (photons) through paired slits one particle at a time. The resulting light pattern is that of a wave or interference pattern.

But how can that be possible, if only one particle went through?

The answer, strangely enough, is the particle was interfering with itself. As the particle traveled, it was traveling as a wave of probability, passing through both (and all other paths) simultaneously.

This was not what was expected and was against popular thought of the time. The experiment was then modified with detectors to allow scientists to know which slit the particle passed through, with a surprising result. The wave interference pattern disappeared, replaced by a spotted scatter pattern.

This is true even if they try setting up the detectors behind the slits. No matter what the scientists did, if they detect the particle "observe", the interference pattern fails to emerge.

And deeper we go.

"If eyes are windows into the soul, books are rabbit holes into the imagination." — Seth King

Scientists tried a variation on the double slit experiment, called the delayed choice experiment. The delayed choice experiment, delays when the particle is detected. But to no avail, even if the particle is detected after the fact, the interference pattern does not emerge. This means (at least in the quantum world) the act of observing can influence events, even if those events have already happened (An Important Clue!).

The act of observing collapses the wave pattern from information which otherwise exists in an undetermined state of probability into the stuff of reality or a single particle.[7]

SCHRODINGER'S CAT PARADOX

How can the cat be both dead and alive? That's Schrödinger's cat paradox. It's a thought experiment that highlights the peculiar nature of quantum superposition. It imagines a scenario where a cat is both alive and dead simultaneously until observed, illustrating the strangeness of quantum mechanics.

Schrödinger's cat paradox is a fascinating idea in quantum mechanics. Imagine a cat in a closed box that can be both alive and dead at the same time. This puzzle makes us question how things work in the quantum world, where something can be in multiple states until observed or measured.

Schrödinger proposed this paradox to question the Copenhagen interpretation of quantum mechanics, which posed conceptual difficulties. While the cat paradox is often cited for its seemingly absurd nature, it carries philosophical implications that stem from the potential consequences of quantum mechanics (Insert laugh track from Big Bang Theory here…).

---

[7] To date, much larger particles and even DNA have been seduced into undetermined states of probability.

It is interesting to note that Schrödinger, the originator of the paradox, did not adhere to a materialistic perspective. Instead, he believed in the existence of a universal Mind that accounted for everything, suggesting a non-materialistic worldview.

From the perspective of the Conscious Anthropic Principle (CAP), the paradox takes on a different interpretation. In CAP, the interaction between collective consciousness and shared fractal information is fundamental to the existence of life. Therefore, the cat, as a conscious entity, can have a say in its own destiny and is connected to the collective consciousness. From this viewpoint, the cat's self-observation and participation in the collective consciousness play a role in determining its state, resolving the paradox.

Schrödinger's cat paradox challenges our understanding of quantum mechanics by showcasing the concept of superposition. However, from a CAP perspective, the paradox can be resolved as the cat, as a conscious being, is not solely subject to external observation but has a role in shaping its own destiny through self-observation and its connection to the collective consciousness.

The Empowering Essence of Questions

As I cautiously ventured out on my amazing book writing caper, I thought I would have a number of Eureka moments. In truth, there really weren't that many.

Eureka means to cry out with joy or satisfaction when one finds or discovers something. But what I found was even more fulfilling — I fell in love with the questions themselves.

I began my journey armed with a little knowledge, which, as they say, can be a dangerous thing. I also had a few haunting mysteries that had been gnawing at my curiosity. Just like the intrepid gang from Mystery Inc., I decided to split up and look for clues. Here are the clues I followed on my intellectual adventure:

1. DNA, the miraculous fractal artifact.
DNA, the intricate code of life, holds secrets that begged to be unravelled. It was like a cryptic puzzle waiting for someone to decipher its meaning.

2. Time, what the heck is time?
Time, the enigmatic river that flows ceaselessly, has perplexed minds for centuries. Questions that demanded answers were its nature, its essence, and its place in the grand scheme of the universe.

After extensive research, inward pondering, and more than a few "Jinkies" moments, I finally stumbled upon the inner workings of electricity. This revelation completely shook my intellectual house of cards, and there were no eurekas to be found. Yet, this absence of a Eureka moment did not deter me. Instead, it opened a new door, revealing a vista of possibilities and hinting at a grander enigma that lay beyond.

This journey led to the formulation of my four pivotal questions:

1. Are we an emergent phenomenon?
The intricate web of existence seemed to imply a deeper interconnectedness, a cosmic puzzle in which each fragment represented information yearning to be unveiled.

2. Is everything a function of consciousness and information?
Consciousness, the elusive essence that defines our existence, stood at the crossroads of information and self-awareness. Understanding their role in the universe became paramount.

3. And, oh, I forgot, what the heck is time?
The age-old question about time keeps drawing us in, hinting that its solution might be hidden in the world of information and consciousness — a query I enjoy posing at dinner parties.

4. Why are we here?
The ultimate question regarding the purpose of our existence loomed large, circling back to all the others in search of a resolution.

Though the mysteries of DNA and time remained unsolved, the first few pieces of the puzzle came together with the revelation of how electricity works. This newfound knowledge not only deepened the mystery but also provided a glimpse into the interconnectedness of the universe.

In this journey, I discovered that the joy of discovery wasn't always encapsulated in a single Eureka moment. Sometimes, the process itself, the pursuit of answers, and the unrelenting curiosity fueled the journey. As I delved deeper into these questions, I realized that they were not just intellectual pursuits but reflections of our collective yearning to understand the universe, our place in it, and the profound mysteries that continue to beckon us forward.

Ultimately, it wasn't about finding all the answers but finding fulfillment within the questions that led to an ever-expanding horizon of knowledge and wonder. Just like Mystery Inc., I continued to chase those clues, knowing that the thrill of the chase was as rewarding as solving the mysteries themselves.

MY FIRST HINT AT A BIGGER WORLD

ELECTRICITY

Electricity is described by the flow of charged particles, usually electrons, through a conductor like a wire, driven by a difference in electrical potential (voltage) and influenced by electric and magnetic fields. This electric current can power devices and perform work.

How does electricity actually work? (Not how you may think).

To truly grasp how electricity operates within an electric circuit, we must first dispel a common misconception. Often, we liken the flow of energy to water coursing through a pipe, using current to represent energy volume, voltage to symbolize pressure, and resistance to denote the narrowing of the conduit. While this analogy offers a straightforward explanation with captivating visuals, it oversimplifies the true nature of electricity and fails to capture the intricate interplay between electric and magnetic fields, which are pivotal for energy transfer in a circuit.

Electricity, an enthralling and indispensable aspect of our modern lives, powers everything from our homes to cutting-edge electronics. However, understanding the flow of electricity necessitates delving into the intricate relationship between electric and magnetic fields, as well as the transfer of energy.

Embark on an exciting journey delving into the intricate landscape of electric energy transfer. As we navigate this captivating expedition, our exploration is destined to circle back to profound questions about the essence of our existence. Along the way, we'll encounter the fascinating concept of fractal information — a dynamic blueprint intricately embedded in every facet of our reality. The goal is to empower you with knowledge, sparking contemplation on whether everything surrounding us is, at its core, a manifestation of information. Are you ready to embark on this illuminating adventure? (Question 2: *Is everything a function of consciousness and information?*)

Think of electricity as a way to transfer energy over distances, like storing power for later use. It's not exactly like water flowing through pipes, as we often imagine. Instead, it involves the interplay of electric and magnetic fields.

Imagine you have a battery or generator as a power source, and it creates an electric field that surrounds it, creating fields that can be thought of as a bubble of energy. When you connect a circuit, something magical happens. Tiny particles called electrons start moving through wires because of this electric field. It's like they're pushed in a specific direction, creating what we call an electric current, and electricity starts flowing.

As these electrons move, they also create a magnetic field around the wires. The stronger the current, the stronger this magnetic field becomes. Now, here's where it gets interesting. This changing magnetic field can induce an electric field in other nearby wires, even if they're far away from the power source. The bubble of energy can even manifest around distant loads. This is how electricity can travel long distances.

Throughout this journey of electrical energy, something important happens: energy is conserved. It doesn't disappear, it transforms into different forms, like kinetic energy (movement), electric potential energy (stored energy), or even heat and light because of resistance.

So, understanding electricity means understanding how these electric and magnetic fields work together to transfer energy. It's like a well-choreographed dance where energy changes forms but never vanishes, even in the vacuum of space, which is a *very* important clue. This stored energy bubble can manifest in a vacuum, which means it is using the canvas of reality as its storage device.

This is where my door to understanding had a train drive through it.

By exploring these concepts, we gain a deeper understanding of how electricity behaves and the elegant laws that govern it. And it raises fascinating questions about the nature of our reality, like whether everything around us is ultimately a manifestation of information as viewed by a conscious observer.

> *"We will make electricity so cheap that only the rich will burn candles."*
> — *Thomas A. Edison*

Quantum Entanglement

This is what reveals a remarkable manifestation of shared information.

Quantum entanglement, observed at the quantum scale, demonstrates how entangled particles maintain a profound connection, allowing for the instantaneous sharing of information, regardless of the physical distance between them.

Within the framework of shared fractal information, quantum entanglement reflects the fundamental principle that information is intricately woven into the fabric of the universe. It suggests that the entangled particles are not isolated entities but rather integral parts of a unified whole, connected through shared patterns of information encoded within the fractal structure of reality.

Quantum entanglement is often called "spooky action at a distance." But I see it as something even more amazing — a miraculous sharing of knowledge closer than you think. It's like discovering hidden connections that tie everything and everyone together. Even though it might sound spooky, it fills me with a sense of wonder and amazement.

Imagine quantum entanglement as a kind of magical connection between particles, like a pair of special phones that instantly share secrets, no matter how far apart they are. This connection shows us that the space between particles is like an illusion when it comes to sharing information.

In this view, it's as if these particles are like pieces of a complex puzzle, where changes in one piece immediately affect the other. This reveals a deep unity and connection that goes beyond what we normally think of as physical boundaries.

From this perspective, quantum entanglement invites us to see the universe as a giant network of interconnected information. Instead of separate things, it's like everything in the universe is part of a big, interconnected system where shared patterns of fractal information are exchanged instantly.

"*Understanding*" quantum entanglement in this way helps us realize that the universe isn't just a bunch of separate pieces. It's a unified and interdependent system where every part communicates through shared patterns of fractal information, kind of like how a symphony works with each instrument playing its part to create beautiful music.

"I think I can safely say that nobody understands quantum mechanics."
— Richard Feynman

Superconductivity

Superconductivity is renowned for its extraordinary ability to conduct electricity without any energy loss.

Superconductivity holds immense potential not only in the realm of technology but also for shedding light on the miraculous functioning of the human brain. Imagine superconductivity as a special key that has the power to turn the ordinary into the extraordinary. This remarkable property, typically found in specific materials at extremely low temperatures, hints at an anthropic key within the complex operations of the human brain. It's as if this special key could unlock hidden doors of understanding, revealing the anthropic secrets of our existence, and help us to gain an understanding of the brain's truly extraordinary power of higher thought.

From an exploratory perspective, the idea that superconductivity may have an anthropic significance and is necessary in the mind of a conscious observer is intriguing. While the direct relationship between superconductivity and brain function remains a subject of ongoing research, contemplating the potential implications is truly awe-inspiring.

One intriguing notion is that superconductivity or something similar, a property enabling electricity to flow without loss, plays a role in explaining why our neural network is so incredibly fast and efficient.

Understanding how superconductivity operates in the brain could lead to significant improvements in how we process and transmit information. This could shed light on why our brains excel at tasks like thinking and connecting different ideas, as well as on the mysterious nature of consciousness.

While we continue to explore and speculate on the exact role of superconductivity in the brain, its possible presence suggests a significance beyond technology. It hints that superconductivity might be a fundamental aspect of how our conscious minds function, playing a crucial role in a universe that requires conscious observers.

Recognizing the profound connection between superconductivity and the intricate workings of the brain not only expands our understanding of how the universe's emergent properties may have influenced the evolution and capabilities of our most complex organ but also prompts us to consider a captivating possibility: that the interplay between physics and cognition could be a reciprocal relationship.

By delving into the relationship between superconductivity and brain function, we catch glimpses of the purposeful design and interdependence between the fundamental fabric of the cosmos and the intricate mechanisms of our minds. This realization provides deep insights into the profound relationship between physics and cognition, offering a glimpse into the possibility of a reciprocal influence that exists between the functioning of our minds and the underlying principles governing the universe.

In this perspective, the remarkable properties of superconductivity may serve as a window into the purposeful interplay between the laws of physics and the emergence of cognition. Simultaneously, this understanding opens up the possibility that our cognition and conscious experience play a role in shaping and interacting with the fundamental fabric of the cosmos.

> *"If you think about consciousness long enough, you either become a panpsychist or you go into administration."* — John Perry

When we investigate the potential connection between superconductivity and the brain, we attempt to gain a deeper understanding of how the universe and our minds are interconnected. This encourages us to ask a significant question:

Can our thinking actually affect and shape the rules that govern our reality?

As we explore this idea, we uncover a fascinating potential perspective of the universe. The study of superconductivity and neuroscience doesn't just lead to technological progress; it can also provide profound insights into the deliberate design and dynamic interplay between physics and thinking.

Quarks

Quarks are elementary particles and fundamental building blocks of matter.

But what are Quarks? They are a type of Lego-like subatomic particle, smaller than atoms and even smaller than the protons and neutrons that make up the nucleus of an atom. Quarks are considered fundamental because they are not made up of smaller particles. They are indivisible and do not have a size or structure like the atoms they are a part of, in other words, quarks are the most basic constituents of matter and are little more than information. Quarks appear in pairs and are held together by what is called the strong force.

The Periodic Table of Elements

This is like a big chart that shows all the different ingredients in the amazing fractal cookbook of this reality.

Imagine the periodic table of elements as a vast pantry in the cosmic kitchen, filled with countless ingredients from the amazing fractal cookbook of our reality. This pantry is cleverly organized, just like a well-arranged spice rack, grouping similar elements based on their unique flavours or personalities.

Each element in the periodic table has its own unique symbol, like H for hydrogen or O for oxygen. Elements are made up of tiny particles called atoms. Atoms are like Lego blocks of matter because they can join together to form different things.

Two important things we can learn from the periodic table are the atomic number and atomic mass of each element.

The atomic number tells us the number of protons in an atom's nucleus. Protons are tiny particles with a positive charge. Each element has a different number of protons. For example, hydrogen has 1 proton, while oxygen has 8 protons. The atomic number determines the identity of an element.

The atomic mass tells us the total mass of an atom, which is a combination of its protons and neutrons. Neutrons are particles in the atom's nucleus that have no charge. Elements can have different numbers of neutrons, but as long as they have the same number of protons, they are still the same element. For example, there are different forms of carbon called isotopes, which have different numbers of neutrons but the same number of protons.

The periodic table helps us understand how the atomic number can influence the properties of atoms, shaping their personality. Elements in the same column of the periodic table, called a group, tend to have similar properties because they have the same number of electrons in their outermost shelves. Electrons are negatively charged particles that frame the nucleus of an atom. The outermost shelf is important because it determines how atoms interact with other atoms. For example, elements in Group 1, like hydrogen and lithium, have 1 electron in their outermost shelf and are very reactive. They easily give away their outermost electrons to other elements.

The periodic table is akin to the ultimate cereal aisle in the grocery store of the universe, with each element serving as a unique ingredient listed on the cosmic cereal box. Just like the back of your favourite cereal box reveals precisely what's inside, the periodic table unfolds the essential components composing everything around us. Each element boasts its own special recipe, distinguished by a unique atomic number and atomic mass.

Envision hydrogen as a dash of one proton, helium as a sprinkle of two protons, and so forth. As you peruse the list, it's akin to navigating through different types of breakfast treats, each box harbouring a surprising blend of elements. The periodic table not only organizes them neatly but also unveils patterns, much like discovering that your favourite cereals are grouped on the shelf.

Now, when you grab a handful of elements from the same row or column, it's akin to choosing ingredients that share similar flavours. They're chemically related, mirroring the difference between crunchy and puffy cereal. The periodic table is your cosmic kitchen guide, revealing which combinations will create a delicious chemistry dish and which might result in a fizzy explosion.

It's a delightful journey through the cosmic cereal aisle, where the atomic number and atomic mass serve as the nutritional labels for each element. As you stroll down the cosmic grocery store, picking up elements like you would your favourite cereals, you're essentially curating a recipe for the universe's grand feast. Just like selecting ingredients for a tasty meal, the periodic table guides you in creating cosmic combinations with harmonious flavours.

In this cosmic culinary adventure, the periodic table highlights our reality's fractal blueprint, unveiling the intricate and interconnected nature of information. Each atomic detail echoes a larger cosmic pattern, mirroring the fractal essence that underlies the fabric of our reality. So, as you explore the universe's cereal aisle, appreciate the fractal dance of information that shapes the cosmic recipe of existence!

ODD FACT

Ever noticed that certain elements are more abundant than others? Apart from hydrogen, the most prevalent of all elements, every other element undergoes a continuous cycle of birth and death intricately tied to the life cycles of stars. The ratio of abundances among elements is shaped by the processes occurring in the cores of stars. Elements are forged through nuclear fusion in stars, where lighter elements are fused into heavier ones. The specific conditions and reactions in stellar cores determine the relative abundance of each element in the universe. This cosmic alchemy, occurring over billions of years, establishes the distinctive ratios we observe among elements.

Magnets

A magnet is an object that produces a magnetic field, which is a force that can attract or repel certain materials. Magnets have two poles: the north pole and the south pole. Similar poles repel each other, while opposite poles attract (opposites attract... or so I've been told).

The concept of permanent magnetism or domain electro is related to the microscopic structure of specific materials. These materials contain tiny regions called magnetic domains, where the atoms or molecules align their magnetic moments in the same direction, resulting in a stronger overall magnetic effect.

In the periodic table of elements, if we look at atomic numbers 26, 27, and 28, corresponding to the elements iron, cobalt, and nickel, we find that these atoms have unique properties. Their relatively small atomic size and the lop-sided arrangement of electrons contribute to these properties. At room temperature, the magnetic domains within these elements naturally align, making them ferromagnetic and capable of exhibiting magnetic properties without the need for external assistance. Because of these characteristics, they are widely used in various applications that rely on magnetism, such as electric motors, generators, and magnetic storage devices.

The periodic table of elements serves as a valuable tool for identifying elements with magnetic properties. It enables scientists and engineers to understand and utilize the distinct characteristics of different materials for practical purposes. By referring to the periodic table, researchers can explore and harness the magnetic properties of specific elements, leading to advancements in technology and various fields of science.
(Please see Blacksmith's Tale for a simple description.)

Spin

Spin is akin to an inherent twist possessed by particles, manifesting in fixed amounts that influence their behaviour in magnetic fields and various quantum physics scenarios. It serves as informative data about particle behaviour.

Spin is a property that particles have, and it comes in quantized values. For particles like electrons, their spin can be either +1/2 or -1/2, while particles like photons can have spin values of +1, 0, or -1.

The terms "up" and "down" refer to the orientation of the particle's spin in a magnetic field. If a particle has "spin up," its spin aligns with the magnetic field, and if it has "spin down," its spin opposes the magnetic field.

In essence, spin is a way of describing the intrinsic angular momentum of particles, and its quantized values and orientations play a role in how particles behave in different situations, especially in the presence of magnetic fields.

Magnetic Field Lines

Back to science class. Take an ordinary bar magnet, place a piece of paper on top, and then sprinkle iron filings onto the paper.

What do you see?

It is an experiment we have all seen before. But I ask you to question:

What are we seeing?

We observe lines of force radiating from the poles of the magnet.

However, what are these lines made of?
Not the iron filings, not the paper, not the magnet itself.

The lines, what are the lines themselves made out of?

Are they composed of electrons, photons, or virtual electrons? Is the magnet emitting energy?

I would suggest that these "invisible lines of force" are manifestations of information and can only be perceived by a conscious observer. Some may consider this notion heretical.

The Peltier Effect
This is a phenomenon where an electric current flowing through a junction of two different conductive materials can either absorb or release heat, depending on the direction of the current.

This effect is utilized in thermoelectric coolers, such as those found in portable refrigerators, to keep drinks cold by using electricity to transfer heat information away from the cooling compartment.

The Peltier effect fascinates me. You know those coolers you plug into your car to keep your beverages cold, right? I was shopping for chillers for the Lab and decided to research the different types available, stumbling upon Peltier effect coolers. I was familiar with them because I "occasionally" use them to keep drinks cool in my car. So, I decided to delve deeper into how they work.

"Never let the truth get in the way of a good story." — *Mark Twain*

Essentially, Peltier effect coolers involve two dissimilar metals. When you pass current through these metals, something intriguing happens. Typically, when you pass current through a wire, energy or information is transferred. There's a difference in electric potential, which can be converted into potential energy, and work gets done. However, in this case, when you pass current through these dissimilar metals, one side gets cold, and the other side gets hot. Now, that doesn't immediately make sense, does it?

To understand this, we have to go a bit deeper, into the realm of atoms. At very low temperatures, atoms sit relatively still. As they get hotter, they start to vibrate, and that's what we call heat. Heat essentially is the vibration of atoms, and cold represents the lack of movement, a slower state. This explains why there's a fundamental limit to how cold something can get — it can only get as cold as when these vibrations cease.

However, when it comes to heat, things can get really, really hot. Here's the thing: it can only get so cold, and what's happening there is atoms vibrating differently. It's like that — what I keep saying is that it's all about information and nothing else. It's not about mass or electron flow, it's purely about information. This is another crucial clue that gets me really excited. I think, *"Wow, this odd beer chiller is essentially transferring information."*

The underlying principle behind the Peltier effect is thermoelectricity, which involves the relationship between electric current and temperature. When an electric current passes through the junction of dissimilar materials, it results in the transfer of thermal energy or behaviour. As a result, one side of the junction becomes cooler while the other side becomes warmer.

ODD QUESTION

Exploring the weight of fully charged batteries and the enigma of hot and cold.

Have you ever pondered whether fully charged batteries are slightly heavier than depleted ones?

This seemingly Odd question actually leads us on a captivating journey into the realm of atoms and the concepts of hot and cold.

First, let's address the battery mystery. Batteries store energy, which they release to power our gadgets. But does this stored energy affect the battery's weight? The answer might surprise you; yes, it does, but the change is incredibly tiny. When a battery is fully charged, it does weigh a fraction more than when it's depleted. However, this difference is so minuscule that it's typically imperceptible in our daily lives.

Now, let's delve into the intriguing world of hot and cold. When we talk about something being hot or cold, we're essentially talking about the movement of tiny particles called atoms. Picture atoms as minuscule, invisible spheres that compose everything around us, from the air we breathe to the objects we touch.

When atoms move swiftly and vibrate energetically, we perceive this as heat. So, when an object feels hot, it means its atoms are in a lively, high-energy dance. Conversely, when atoms move sluggishly and vibrate less, we sense cold. Cold is like the absence of this atomic activity.

Here's a fascinating twist: remarkably, a hot atom actually weighs ever so slightly more than a cold one. This might sound astonishing, but the difference is incredibly subtle. As atoms gain energy and heat up, they absorb an almost infinitesimal amount of extra mass. This phenomenon is elegantly explained by Einstein's theory of relativity, one of the most profound principles in physics. It tells us that energy and mass are intimately intertwined, and even the tiniest amount of energy can impart a minuscule increase in mass.

Now, concerning cold, there exists a remarkable point known as absolute zero. At absolute zero, atoms come to a virtual standstill, vibrating as minimally as theoretically possible. It represents the coldest temperature anything can ever reach, a frosty -273.15 degrees Celsius (-459.67 degrees Fahrenheit). At this extraordinary coldness, atoms possess the least energy and, intriguingly, the lowest conceivable mass.

So, while the weight change of batteries due to their charge level is incredibly slight, the phenomenon reminds us of the astonishing connections between energy, mass, and the intricate world of atoms that silently shapes our everyday experiences.

Quantum Fluctuations

These are super small, quick changes in energy that happen because things are uncertain at the tiniest level. They're like the flickering of a tiny light in the quantum world.

Quantum fluctuations refer to the temporary and random changes in the amount of energy at a specific point in space, as described by the uncertainty principle in quantum physics. This phenomenon suggests that even in a seemingly empty vacuum, there are fluctuations in energy levels (I believe you can substitute information for energy and gain a clearer view).

According to the uncertainty principle formulated by Werner Heisenberg, there is a fundamental limit to how precisely we can measure certain pairs of physical properties, such as position and momentum. This principle also implies that the energy of a system cannot be precisely known at a given moment in time. Therefore, in the realm of quantum mechanics, energy levels can briefly fluctuate or change unpredictably.

In simple terms, quantum fluctuations mean that particles and their associated energy can spontaneously appear and disappear in a small area of space, as long as these fluctuations do not violate the overall rules and conservation of energy in the larger context.

Remember, time can be thought of as a representation of causality, through the eye of a conscious observer. Time is an integral part of our reality, connected to the expansion of the universe. It is not a separate dimension but rather an emergent construct that emerges from the movement of the present moment to the next moment. As the universe expands, time moves forward (more on this in Question 3).

Think of quantum fluctuations like bubbles that randomly appear and disappear in a fizzy drink. These bubbles represent temporary changes in energy in a tiny space. They happen because the rules of the fizzy drink world allow for this uncertainty.

In the quantum world, particles and energy can do something similar — pop in and out of existence temporarily, following certain rules, and without breaking the laws of physics. These fluctuations are like the fizzy bubbles of the quantum universe, and they're important for things like creating particle-antiparticle pairs and even for what happened in the very early moments of the universe.

(Please note: Fizzy drink world exists in the same quantum state as Margaritaville;)

Water

Water is transparent because if it were not, life, and you and I, would not be possible (anthropic principle).

Water is transparent because of its molecular structure and the way it interacts with light. The transparency of a substance depends on how it absorbs and scatters light. In the case of water, its transparency is primarily due to its molecular composition and arrangement. Water molecules consist of two hydrogen atoms bonded to an oxygen atom, forming a V-shaped molecule. This molecular structure leads to a relatively symmetrical distribution of electrons, which results in water having a relatively low absorption of visible light.

When light interacts with a water molecule, it can be absorbed, transmitted, or scattered. In the case of water, it absorbs light weakly in the visible spectrum, particularly in the red part of the spectrum. This weak absorption allows most of the visible light to pass through the water, making it appear transparent.

ODD FACT

Water possesses its characteristic blue colour because the blue wavelengths of sunlight are scattered in all directions by the water molecules. This scattering phenomenon arises from the interactions between water molecules and sunlight.

Furthermore, water molecules are relatively small compared to the wavelength of visible light, which reduces their ability to scatter light. Scattering occurs when light interacts with particles or irregularities in a substance and changes direction. In the case of water, the small size of its molecules limits scattering, contributing to its overall transparency.

From an anthropic perspective, the transparency of water can be attributed to the specific conditions necessary for life to exist and thrive. The conscious anthropic principle considers that the universe is not a random collection of particles and energy, but rather a purposeful creation that fosters the development of intelligent life.

Think of water's transparency like a special window. This window is crucial for all kinds of life, including us humans, to see and interact with the world around us. It's like having clear glasses that let us see things properly.

Now, imagine if this special window wasn't clear but completely blurry or dark. That would be a big problem. We rely on our sense of sight to survive and get around. And it's not just us, many plants and other creatures use this window too.

If water were not transparent, it would be like having foggy glasses all the time. Seeing things and using sunlight to make energy, which is super important for many living things, would become really tough. This would mess up the whole system of life on Earth, from the tiniest plants to the biggest animals.

So, the fact that water is transparent is like a purposeful design in our universe. It's there to make sure intelligent beings like us can see the world, use light for energy, and be part of the complex web of life on our planet. It's like having the perfect glasses to enjoy the show of life on Earth.

Antimatter

Anitmatter isn't just the stuff of sci-fi starships like the Enterprise, it's a captivating piece of the cosmic puzzle. Often seen as the "evil twin" of normal matter, antimatter is intriguingly aligned with the same fractal blueprint of information that underpins the entire universe. However, it comes with a twist — a built-in self-destruct mechanism. This unstable nature ultimately leads to the scarcity of antimatter in our universe, a phenomenon that raises questions about its purpose. Thanks, "evil" Spock, for this cosmic clue!

In our exploration of reality, we have come to understand that the fundamental constituents of matter, such as electrons and protons, are composed of the same fundamental parts. They possess identical mass and spin, leading us to believe that all subatomic particles are constructed from a limited number of basic components.

This consistency suggests the existence of a universal template or blueprint, represented by intricate fractal patterns, that guides the construction of everything in the vast expanse of the universe. It appears that there is a predetermined structure, reminiscent of a prefab or cookie-cutter approach, rooted in the fractal blueprint of the cosmos. This points to the notion that information holds a universal significance in shaping our reality.

Does the concept of shared fractal information point to a profound interconnectedness and purpose in the fabric of the universe?

Could the existence of a universal blueprint and the presence of shared patterns suggest that information isn't just prevalent but also a foundational aspect of reality, weaving the intricate tapestry of our universe?

THE LIVING

The Domain of the Living

Life is special; don't let anyone tell you it's just a random inevitability. We naively accept that life is normal and abundant.

We live in an absurd universe, but it is that very life that offers our only chance to achieve something worth living for.

Imagine a brave space explorer embarking on a daring journey to an alien planet in search of an extraordinary ancient technology. This technology, beyond the realm of science fiction, possesses remarkable capabilities: it can self-replicate, repair itself, and construct a wide array of objects. Astonishingly, this incredible technology is not a product of advanced machinery or artificial intelligence — it is DNA.

DNA

Short for deoxyribonucleic acid, is a miraculously intricate molecule.

In fact, a single DNA molecule is believed to encompass just over a few billion atoms. Within its graceful structure resides an enormous wealth of information. However, DNA transcends being merely a chemical compound — it is the living embodiment of information itself.

Once inside a living cell, DNA unleashes its potential, serving as a blueprint for life. This fractal information, encoded within the DNA molecule, orchestrates the emergence of complex living organisms. It guides the formation and function of cells, tissues, and organs, culminating in the vast tapestry of life that surrounds us.

The journey of DNA from a minuscule molecule to the grandeur of the living world is a testament to the power of information and its purposeful manifestation. It is through the intricate dance of DNA that life finds its expression, evolving and adapting to its surroundings over eons of time.

The significance of DNA extends far beyond its role in individual organisms. It serves as a universal language, connecting all living beings through shared patterns and codes. From the tiniest microbe to the majestic giants of the natural world, DNA unites us in a web of interconnectedness, reminding us of our shared origins and common destiny.

As our intrepid space explorer uncovers this ancient technology, they come face to face with the awe-inspiring realization that DNA is not merely a molecule, but a conduit for the intricate dance of life itself. It is a testament to the vast complexity and purposeful design inherent in the fabric of the universe.

The remarkable properties of DNA highlight its significance as a repository of information and a catalyst for life. From its self-replicating abilities to its capacity for constructing diverse biological structures, DNA stands as a testament to the beauty and intricacy of the natural world. As we continue to explore the mysteries of the universe, let us marvel at the profound role of DNA, the ancient technology that continues to shape and define life as we know it.

SIDE QUEST

Levinthal's paradox, in simple terms, is a puzzle in the field of protein folding. It highlights the apparent contradiction between the incredibly vast number of possible ways a protein can fold and the relatively short time it takes for many proteins to adopt their correct functional shapes.

To break it down:

Proteins are molecules made up of long chains of smaller building blocks called amino acids. The way these chains fold into a three-dimensional structure is crucial for their function in the body.

The number of possible ways a protein can fold is astronomically large. Imagine having a long, tangled piece of string and trying to arrange it into a specific, complex shape. There are an almost infinite number of ways to do this.

Yet, proteins often fold into their correct shape quickly. This happens despite the mind-boggling number of possible folding arrangements. Many proteins fold correctly in a fraction of a second.

Levinthal's paradox essentially asks, "How do proteins find their correct shape so quickly when there are so many possible ways to fold?" It's a paradox because, based on pure randomness, it would take much longer than the actual time it takes for proteins to fold correctly.

The paradox highlights the fact that there must be some kind of guiding or efficient process at work in protein folding that helps them find their functional shape rapidly. Researchers in biology and biochemistry have been exploring this puzzle for many years to better understand the mechanisms involved in protein folding.

Life, with its extraordinary complexity and remarkable ability to evolve, raises intriguing questions about the origins of its knowledge and behaviour. It appears as though living organisms possess an innate capacity to access information from seemingly unknown sources. This enigmatic quality aligns with the concept of Morphic Resonance, a theory that suggests memory is inherent in nature. According to this theory, natural systems inherit a collective memory from all previous entities of their kind, shaping their development and behaviour.

"The butterfly is a flying flower, the flower a tethered butterfly."
— *Ecouchard Le Brun*

The functioning of living cells and their astonishing ability to perceive their surroundings and respond accordingly is a captivating subject. These microscopic entities can be likened to incredibly efficient 3D factories, continuously processing, and exchanging information at astonishing speeds. Their ability to recognize their location, understand their purpose, and adapt their behaviour accordingly is nothing short of mystifying.

This intricate interplay of information and cellular behaviour lies at the heart of the awe-inspiring phenomenon of life. It prompts us to question the underlying mechanisms that enable living systems to acquire knowledge and navigate their complex environments. The theory of Morphic Resonance offers a thought-provoking perspective, proposing that this inherent wisdom and responsiveness are not mere chance occurrences but rather a manifestation of a collective memory woven into the very fabric of nature itself. By delving into the depths of these mysteries, we may gain a deeper understanding of the profound interconnectedness and purpose inherent in the web of life.

Imagine a music box player, an intricate machine capable of producing beautiful melodies. But what if this player had a consciousness of its own? It could tap into a vast reservoir of shared knowledge, drawing inspiration from the collective memory of all music boxes that have ever existed. Similarly, nature itself seems to possess a set of shared tools, a common pool of information that organisms can access and build upon. This notion hints at a deeper interconnectedness and suggests that life is not an isolated phenomenon but rather part of a greater tapestry.

The emergent process of life can be seen as a journey toward observation, an intricate dance that unfolds through the interplay of countless life forms. Each organism, in its own unique way, contributes to the richness and depth of conscious experience. It is through this vast hierarchy of related life forms working together that intelligent life can flourish, enabling a fuller exploration and expression of consciousness.

This journey of life and consciousness is remarkable in its defiance of entropy, the inevitable tendency towards disorder and randomness described by the Second Law of Thermodynamics. Life, in its complexity and organization, stands as a counterpoint to this law, as it creates and sustains intricate structures and processes. It harnesses energy and information, constantly evolving and adapting to its surroundings. In doing so, life appears to transcend the boundaries of mere matter and energy, displaying an inherent drive towards organization and purpose.

The relationship between life and entropy raises profound questions about the nature of the universe itself. The very fact that life exists and flourishes in a universe governed by the Second Law of Thermodynamics suggests that life and consciousness are not accidental outcomes but integral components of the cosmic fabric. The universe seems to be intricately designed to support the emergence of life, presenting a tantalizing glimpse into a deeper order and purpose that transcends our current understanding.

The enigmatic nature of life's ability to draw information, the concept of Morphic Resonance, and the universal aspects of human experience all point towards a deeper interconnectedness and a shared foundation of knowledge.

Life's journey towards observation and its defiance of entropy further emphasizes the profound role that consciousness plays in the fabric of the universe. As we continue to explore the mysteries of life and the universe, let us embrace the wonder and complexity that surround us, and seek to unravel the deeper truths that lie within our collective existence.

Love makes the world go around.
That & low entropy. — Odd

(More details of life's impossible defiance of entropy in Question 4.)

THE BIG, CLASSICAL PHYSICS AND GENERAL RELATIVITY

The Expanse of the Large

I'd like to put forward the idea that we exist in a perceived matrix of reality with length, width, height, and time as emergent constructs. However, comprehending this concept can be challenging without a solid foundation of knowledge. Just like any substantial proposal, it's essential to examine the validity of our information.

The universe is unimaginably large. The universe we see is only the exceedingly small tip of the unimaginably vast cosmic iceberg. From where we sit, we can see only about 4% of the known universe and that percentage will change very little in the coming centuries. Almost everything we believe we know is an assumption.

The farthest light we can see is called the cosmic microwave background (CMB). It took more than 13 billion years to travel to us. Imagine it's like watching a really old movie on TV.

Now, you might think that means the universe is 26 billion light-years wide, right? Well, here's the twist! Because the universe is stretching out, like when you have a piece of dough and it starts expanding, it's actually closer to 46 billion light-years wide now. So, the universe is incredibly big, and it keeps getting bigger because of this expansion.

Gravity

Gravity is what mass does.
Gravity is the guiding hand of the huge. Gravity is the Architect of form in the Universe.

Gravity shapes the cosmos and is a domain of immense scale and significance. From the grandeur of celestial bodies to the delicate dance of atoms, gravity serves as the master architect of form in the universe.

When we venture beyond the realm of the small, where Quantum Mechanics reigns, into the world of the vast, governed by Classical physics and general relativity, where celestial objects like planets and stars hold sway, we encounter the immense influence of gravity. Here, Einstein's equation $E=mc^2$ comes into play, revealing the close link between energy and mass.

Nevertheless, bridging the gap between this vast gravitational world and the quantum realm of minuscule particles is akin to attempting to connect two different types of puzzle pieces. The rules that govern these two domains don't quite align, and scientists are diligently labouring to discover a unified theory capable of explaining both.

It's somewhat analogous to search for a single instruction manual that can be applied to constructing both a towering skyscraper and a tiny treehouse. This quest for unity represents one of the most significant puzzles in the field of physics!

In the vast expanse of space, gravity reigns supreme. It binds galaxies together in colossal clusters, orchestrating their majestic movements across the cosmic stage. Massive black holes — gravitational behemoths — exert their powerful influence, distorting our spacetime and bending light. Stars, suspended by the delicate equilibrium between gravity and the fiery fusion within their cores, illuminate the heavens with their radiant glow.

On a more intimate scale, gravity shapes the intricate fabric of planetary systems. It moulds planets into spherical bodies, sculpting their landscapes and dictating their orbits. Mountains rise, valleys deepen, and oceans gather under its watchful eye. With a steady hand, gravity maintains the harmony of celestial bodies, ensuring their stability and endurance.

Yet, gravity's influence extends even further, penetrating the realm of the infinitesimally small. At the subatomic level, it wields its power, bringing particles together and forging the building blocks of matter. Within the atomic nucleus, gravitational attraction draws protons and neutrons close, allowing the delicate balance between the strong and weak nuclear forces to maintain atomic stability. Without gravity's guiding hand, atoms, the very essence of our physical existence, would cease to exist.

Moreover, gravity's domain stretches across the vastness of time. It shapes the evolution of galaxies, molding their structures and guiding their growth over billions of years. It dictates the destiny of stars, determining their lifespan and eventual fate. Gravity's unyielding pull orchestrates the dance of celestial bodies throughout the ages, weaving a narrative that spans the cosmic tapestry.

In this grand symphony of creation, gravity emerges as the architect of form. It shapes the universe, from the smallest particles to the largest structures, with its invisible embrace. It sculpts matter, influences energy, and bends the very fabric of our spacetime. Gravity is the architect of undeniable power and elegance, guiding the cosmic ballet with precision and grace.

"Gravity explains the motions of the planets, but it cannot explain who sets the planets in motion." — Isaac Newton

As we gaze upon the wonders of the universe, let us marvel at gravity's domain, for it is the source that gives rise to the awe-inspiring beauty and complexity we behold when we look up to the night sky. From the majesty of galaxies to the intricacies of atoms, gravity stands as the silent architect, shaping the universe and illuminating the profound interconnectedness of all things.

SIDE QUEST

Einstein's iconic equation, $E=mc^2$, provides a captivating glimpse into the complexities of physics. Picture energy (E) as the dynamic currency of the universe and mass (m) as a form of frozen assets within matter. Now, envision the speed of light (c) as the "cosmic conversion rate" — a universal constant governing the transformation of mass into energy and vice versa. In simpler terms, this equation illustrates that energy and mass are interconnected, acting like two sides of the same cosmic coin. It proposes an intriguing idea: the potential for mass to convert into energy, and vice versa, unveiling a profound dance between these fundamental elements.

To convey this concept, imagine a piggy bank as a metaphor for this cosmic exchange. Here, mass represents the coins securely stored within, symbolizing the latent potential for energy. Just as shaking the piggy bank represents a high-energy event, releasing coins (energy), the analogy vividly portrays the conversion of mass into energy. Conversely, collecting scattered coins and placing them back into the piggy bank mirrors the conversion of energy back into mass, representing a lower-energy state. Einstein's equation beautifully encapsulates this extraordinary relationship, showcasing a dynamic ballet between matter and energy in the grand cosmic theatre.

The Universe Is Expanding

Now, when we talk about the expansion of the Universe, we're not saying things are moving within space. Instead, space itself is stretching over time. Imagine space as a rubber sheet getting bigger, causing everything on it to spread out. But here's the important part, there's no "outside" to look at this from. It's like being on a balloon that's getting larger — no matter where you are on the balloon, everything around you seems to move away. This expansion isn't like a regular explosion, it's happening everywhere in the Universe, not just in one place. It's a property of the entire Universe, and it's a concept tied to Einstein's general theory of relativity, which is quite different from our everyday experiences.

In 1912, Vesto Slipher introduced the idea of an expanding Universe. Astronomers noticed something cool when they looked at faraway galaxies — a phenomenon called "redshift." It's like how the pitch of a passing car changes. Redshift happens when things in space move away from us.

Edwin Hubble, an American astronomer, and Georges Lemaître, a Belgian physicist and priest, said that this redshift means the universe is expanding. Lemaître went further, saying if we rewind this expansion, we'd find a time when everything in the universe was super squished together, like a "primeval atom." This idea gave birth to our understanding of space and time.

From an anthropic perspective, I believe the expanding universe highlights the intricate relationship between cosmic phenomena and the conditions necessary for the existence of sentient life. The expansion, with its fundamental role in shaping the fabric of space and what we perceive as time, provides a backdrop for the evolution and emergence of life within the cosmos. From the grand scale of galactic clusters to the minute interactions of particles, the expansion of the universe stands as a testament to the remarkable interplay between cosmological principles and the emergence of conscious observers.

The Universe and the Human Brain

There is a striking resemblance between the universe and the human brain.

This is perhaps more than just an intriguing concept that captivates our imagination. The universe and the human brain share some surprising similarities, even though they're very different in size.

Imagine the universe as a giant puzzle with galaxies scattered everywhere. Now, picture the brain as another puzzle, but this time it's made of tiny cells called neurons, and they talk to each other using special signals.

What's fascinating is that both puzzles, the universe and the brain, are incredibly complex and have lots of connections. In the universe, galaxies are connected like a big cosmic web, kind of like how neurons in the brain are linked.

In the brain, neurons talk to each other with electrical signals and chemicals. In the universe, galaxies are connected by gravity, which helps shape everything.

Another cool thing is that both the universe and the brain can organize themselves and create patterns that repeat, like when you look at a leaf closely, and it has little patterns that look like the big tree it came from. These repeating patterns are called fractals.

But here's the interesting part, when you zoom in and look at both the universe and the brain at a certain scale, they kind of look the same. It's like finding two different jigsaw puzzles that strangely match in one tiny corner.

But remember, even though they look similar at that specific scale, they're still very different and unique. We shouldn't get carried away and start thinking they're exactly the same because they're not.

Exploring these similarities is like going on an exciting adventure that reminds us how everything in nature is connected through these repeating patterns. It's like finding hidden clues in a mystery novel that make us want to learn more about the universe and our brains, which are both incredibly fascinating in their own ways.

I may give you the impression that I am anti-science or a Luddite; however, the truth is that I deeply appreciate the remarkable speed at which knowledge has been accumulated, leading us toward a critical mass of understanding. Standing on the shoulders of intellectual trailblazers, both past and present, we have the unprecedented opportunity to comprehend and enhance every aspect of our existence on an immense scale.

As a species, we find ourselves immersed in a world that is still largely mysterious to us. Like children believing in Santa Claus, we trust in the intricate performance unfolding before us, hoping that it all makes sense. Yet, beneath the surface, there is a realization that this performance, along with our lives, is part of a much grander plan that has been meticulously crafted over billions of years.

Let's continue to summarize and build our understanding:

- The Earth was formed 4.5 billion years ago.
- DNA is an extraordinarily complex molecule and contains all the necessary ingredients for the emergent process of life (EPL).
- Life has emergent properties.
- Life was observed on Earth as soon as it was possible. Instantaneously (give or take a million years...).
- We only have one tree of life. All life on Earth is descended from this one occurrence and in the last 3.8 billion years it has never happened again.
- Successful life has more offspring.
- The emergent process of life is moving towards observation.
- Intelligence allows life to have a richer and fuller experience of consciousness.
- Human beings seem to have free will or choice.
- The numerous tangible forces, rules, and principles that shape our reality seem to be manifestations of applied information.
- The universe we observe seems perfect for us, which implies an anthropic bias.
- The universe seems to be built with an evolving blueprint of fractal information and that information can be shared.

BREADCRUMBS

I wanted to make the emergent event of Right Now more personal because it truly is deeply personal. The emergent event we find ourselves in, is not some distant abstraction, it's the emergent event of the present moment. It's actually intensely personal. I may have unintentionally conveyed it as distant or impersonal, but it's happening right now. We're feeling it. We're experiencing it. And its significance may be even greater than we realize, right?

Perhaps I need to frame this in a more personal context because, in reality, it is personal. I believe this is how we sometimes lose our perspective, just like how we can lose sight of gravity.

Gravity is something so fundamental that it becomes almost invisible. We grow comfortable with not knowing or not fully appreciating it.

You might wonder why I often mention gravity, almost to the point of obsession. The reason is that gravity is unlike anything else; it's unique. It distorts not just space but also time, and it's one of the few forces that truly affects everything in the Universe. I see it as a clue, a glimmer of understanding. Examining gravity closely and truly grasping its nature could potentially unlock a deeper comprehension of everything else.

It's almost like breadcrumbs in a dense forest — they're always right there in front of us, yet we're not quite sure what they signify. Gravity consistently beckons us to ask questions to delve deeper into the mysteries of the Universe, and hopefully, it can lead us to a more profound understanding. That's why I keep coming back to gravity; it's not just an answer in itself, it's a question. Gravity is a question. Gravity is an enigma, and its influence ripples through the fabric of the cosmos. If we question it rigorously, we might find ourselves on the path to a better understanding of the Universe.

Information and consciousness form the very framework of our reality. We awaken each day to view the world open in front of us unaware of how accurate that metaphor is.

SIDE QUEST

In his book *The Goldilocks Enigma* (2006), Paul Davies explores various perspectives on the nature of the universe. I will now summarize these perspectives and offer intriguing insights into our existence and the underlying principles that shape our reality:

1. The absurd universe: This viewpoint suggests that our universe's characteristics and existence are purely coincidental. There is no deeper meaning or purpose behind its formation.

2. The unique universe: According to this perspective, there is a profound unity in the field of physics that mandates the precise configuration of our Universe. It implies that a comprehensive Theory of Everything will eventually clarify why the universe exhibits the specific features and values we observe.

3. The multiverse: This concept posits the existence of multiple universes, each with its own unique combination of characteristics. Within this framework, it is inevitable that we find ourselves in a universe that allows the existence of life, given the vast array of possibilities across the multiverse.

4. Intelligent design: This perspective asserts that a conscious creator intentionally designed the universe to support complexity and foster the emergence of intelligent life. It implies that the fine-tuned conditions we observe are a result of deliberate intentionality.

5. The life principle: According to this principle, there is an underlying cosmic principle that drives the evolution of the universe towards life and mind. It suggests that the emergence of life is not merely a random occurrence but an inherent tendency of the universe.

6. The self-explaining Universe: Wheeler's Participatory Anthropic Principle (PAP) proposes a closed explanatory or causal loop, suggesting that perhaps only universes with the capacity for consciousness can exist. This viewpoint implies a reciprocal relationship between consciousness and the existence of the Universe.

7. The fake universe: This intriguing concept speculates that our reality is actually a simulated virtual reality rather than a physical universe. It proposes that we are inhabitants of an artificially created simulation.

Each of these perspectives provides a unique lens through which we can contemplate the nature and origins of our Universe. They spark thoughtful reflection and promote further exploration into the fundamental questions of our existence. My contribution to these perspectives includes CAP and its variants.

8. The Conscious Anthropic Principle (CAP) offers a fascinating perspective on the nature of the Universe. According to CAP, the fundamental mechanism allowing life to exist is the interaction between collective consciousness and shared fractal information. This principle suggests that consciousness precedes the formation of the universe itself, indicating a conscious-first perspective where reality is self-emergent.

In contrast to the idea of a random assortment of particles and energy, CAP proposes that the universe is a contemplative creation intentionally fostering the development of intelligent life. It challenges the notion of mere coincidence and emphasizes the purposeful nature of the universe, driven by conscious intentionality. Within this framework, consciousness interacts with shared patterns of fractal information to shape the universe, establishing the finely tuned conditions we observe.

CAP invites us to acknowledge that the universe is not solely a product of chance but a deliberate construction imbued with conscious intent, shaped through shared fractal information. It encourages contemplation of the profound interconnectedness between consciousness, the universe, and the emergence of intelligent life. This perspective invites further exploration and inquiry into the mysteries of our existence. (CAP is a sub-variant of Cosmopsychism).

Cosmopsychism, another perspective, offers a unique viewpoint, akin to a top-down gaze at the universe's consciousness. It suggests that the cosmos is a singular, vast sentient being, much like a single gem with many facets contributing to its brilliance.

In this outlook, universal consciousness acts as the core of this multifaceted gem, influencing the consciousness of individual entities, akin to streams of awareness flowing from this cosmic gem. Each facet, representing living beings and inanimate objects, contributes to the greater cosmic mind. This analogy underscores the profound interconnectedness of all existence with universal consciousness, fostering a deeper appreciation for the unity of the cosmos.

Where the Trail Leads

I am suggesting that Universal Consciousness and Shared Fractal Information are not just abstract concepts, they are like beacons that guide us through the enigmatic fabric of existence, leading us with expanding questions.

In various spiritual and philosophical traditions, Universal Consciousness, often referred to as Cosmic Consciousness or the One Mind, takes a place of reverence. This idea whispers of a singular, all-encompassing consciousness that forms the very foundation of all that exists. It's as if a collective awareness transcending the limits of time, space, and individuality wraps around us like a warm embrace.

Universal Consciousness gently nudges us to recognize that we aren't isolated entities but integral fragments of a grander whole. This perspective, like a comforting hug, invites us to ponder the fluidity of the boundaries that separate us from the world, hinting at a profound interconnection with everything and everyone around us.

On the flip side, the concept of Shared Fractal Information brings a sense of marvel. It hints that the universe follows intricate, self-repeating patterns, much like the mesmerizing patterns of a snowflake or a seashell. These intricate structures don't just end at the surface, they extend deep into the heart of information itself.

Shared Fractal Information suggests that the information orchestrating the universe's dance adheres to discernible, recurring designs. These blueprints, like ancient secrets unveiled, guide the formation and organization of everything, from galaxies sprawling across the cosmos to the delicate veins tracing a leaf's surface. It's as if the universe speaks a timeless, consistent language, a language filled with wonder.

The captivating relationship between Universal Consciousness and Shared Fractal Information proposes that the universal consciousness holds within it the very essence of the universe's secrets. In simpler terms, it cradles the fundamental templates of information that breathe life into the cosmos.

Imagine Universal Consciousness as an eternal wellspring of wisdom and awareness, a sanctuary where the essence of existence dwells. Alongside this vast consciousness lies the blueprints, the fractal information, that gently guide the unfolding of the universe's intricate masterpiece. It's akin to Universal Consciousness being the guardian of the "master plan," orchestrating the structure, evolution, and interconnectedness of everything that exists. This realization fills me with a sense of reverence and awe.

This interplay suggests that our individual existence emerged from within consciousness, and that flicker of awareness we are now observing is a shared intimate bond with that same universal consciousness. We are not distant from this cosmic source, instead, we are its living expressions of it, like stars in the night sky contributing to the grand cosmic spectacle. As we embark on journeys of self-discovery and expand our understanding of the world, we tap into the universal wisdom and contribute to the ever-evolving fractal narrative. This idea, this sense that we are all witnesses to the divine resonates with me and makes me feel both humbled and empowered.

The Poetry of Mankind:
An Emergent Dance of Fractal Information and Universal Consciousness

In the grand tapestry of the Universe, humanity stands out as an extraordinary manifestation of intricate patterns and limitless potential. Our presence can be viewed as a poetic dance that intertwines with fractal information and the universal consciousness that forms its foundation.

Picture a snowflake, a tiny crystal born from water vapour. At first glance, it may seem fragile and simple. Yet, as we zoom in, we uncover a captivating complexity. Every snowflake reveals a one-of-a-kind, intricate design — a fractal. This mirrors how our lives unfold, appearing straightforward on the surface but unveiling layers of profound meaning and connection upon closer examination.

Fractals are patterns of information that repeat at various scales, and they permeate the natural world. Our human experiences also exhibit this fractal nature. From the bonding of atoms to create molecules, to the branching of trees, and even the structure of our neural networks, fractal patterns are omnipresent. As a species, we are participants in this magnificent fractal dance, intricately entwined with the patterns of the cosmos.

But what animates these patterns, elevating them beyond mere mathematical curiosities? It is the universal consciousness — the fundamental essence that unites all existence. Think of it as the life force coursing through the veins of the Universe, connecting every atom, star, and living being. It empowers us to perceive, think, and feel.

As humans, our consciousness represents a distinct expression of this universal consciousness. We are akin to individual notes in a cosmic symphony, each contributing a unique melody to the grand composition of existence. Our thoughts, emotions, and actions send ripples through the fabric of the Universe, fostering a harmonious interplay of experiences.

Consider the way we connect with one another. We form relationships, share stories, and build communities. These interactions resemble the intricate interlocking patterns of fractals, with each connection enhancing the depth of our lives. We are not solitary entities but rather integral components of an emerging network of consciousness — a splendid tapestry of shared experiences.

Our capacity to create and express ourselves through art, music, and literature serves as a testament to this emerging nature. Artists draw inspiration from the universal consciousness, channelling it into their creations. When we admire a painting, listen to a song, or read a poem, we are touched by the universal essence within the specific, linking us to a profound, shared human experience.

Within the poetry of humanity, we find a delicate equilibrium between the microcosm and the macrocosm, between the individual and the universal. We are the narrators of our own life stories, crafting verses filled with meaning, love, and discovery. We embody the fractal manifestations of a cosmos overflowing with beauty and awe.

By embracing this perspective, we discover comfort and inspiration in the interconnectedness of all existence. We are not separate from the Universe, we are an essential part of it. Our lives represent the poetry of emergence, a dance merging fractal information and universal consciousness, where each of us contributes a unique and irreplaceable role in the grand cosmic narrative.

AND, OH, I FORGOT, WHAT THE HECK IS TIME?

Question 3

The timeless question about time continued to beckon, its answer perhaps nestled within the fabric of information and consciousness.

"Do we exist in time, or does time exist in us?" — Carlo Rovelli

In the vast expanse of time, significant events have shaped our existence. 13.7 billion years ago, the universe burst into being, setting the stage for the unfolding cosmic drama. Earth joined the cosmic stage 4.5 billion years ago with its remarkable attributes. Then, roughly 300,000 years ago, the age of humanity commenced, marking our unique journey as conscious beings.

Consider the enormity of time. When compared to the total duration, the age of man represents less than 0.00006% of this expansive timeline. Our individual lives, if fortunate, span around 100 years, which accounts for a mere 0.000000022% of the entire duration.

Now, let's compare this against the chances of winning the lottery. Winning the lottery is often seen as a stroke of luck, a rare event with odds heavily stacked against us. The probability of winning depends on various factors, such as the number of tickets sold and the specific lottery rules. On average, the chances of winning the lottery are relatively small, with odds like 1 in 14,000,000 for example.

When we contemplate the immense expanse of time and the short moments we have in our lives, it becomes clear that anyone alive and able to observe this beautiful world today is very fortunate.

This realization prompts us to reflect on the preciousness of our existence and the opportunities we have. While the probability of winning the lottery may seem quite small, the chances of experiencing the wonders of life, forging meaningful connections, and positively impacting the world are within our grasp.

Instead of solely fixating on the slim odds of winning a game of chance, let us embrace the grander perspective that time offers. Let us cherish the moments we have, make the most of the time we are given, and strive to create a meaningful and fulfilling journey.

In the vast cosmic dance, our lives may be but a fleeting moment, yet within that moment lies the potential for profound experiences and contributions. As we navigate the enigmatic currents of time, may we recognize the inherent value of our existence and seek to make each passing moment count.

But what is time?

Time is memories of warm beaches, springtime, and old friends who were once young.
Time is lunch breaks and paychecks.
Time is our hope for our children and our optimism for the future.
Time is a reflection of a lifetime.
Time is an enigma, and time is a question.

I believe that the present moment of "Now" is a clue, an enigma, a question. Now is our window of opportunity.

The Stopwatch

On several occasions, I've found myself engaged in discussions about time at parties and social gatherings. This isn't my usual course of action, yet I enjoy delving into philosophical musings under the right circumstances. I'm genuinely intrigued by people's perspectives and even more so by kindling their own queries. Time is a concept woven into our lives, akin to birthdays, weddings, births, and deaths.

When I ask individuals what time truly is, their reactions are often perplexed, their responses delayed. Most typically resort to describing time as minutes, hours, or days — the components we designate as time.

But my query goes beyond that, I inquire about the essence of time itself. This usually triggers another thoughtful pause, followed by inquiries such as, *"What do you mean, time is time..."* Some may venture that time is the fourth dimension, to which I clarify that time is not a dimension.

If it were, we could measure the past, which we cannot.

The past, I explain, exists solely in the recesses of our minds, dwelling as memories and thoughts. Similarly, the future only takes form within our minds through extrapolated information and probabilities. Our sole portal into the realm of time is the present moment.

To illustrate, I often invoke the stopwatch analogy: if we initiate a stopwatch by clicking "start," that very act signifies the present. After waiting 15 seconds and glancing at the stopwatch, we're no longer observing the past, rather, we're witnessing the present, precisely 15 seconds into the progression of time. The passage of half a minute does not alter this, we're still encapsulated in the present. Even after a full minute elapses, stopping the stopwatch leaves us with a reading of the present.

People are usually left speechless, contemplating inwardly and taken aback by this unusual perspective on an everyday wonder that clearly went unnoticed before.

Imagine an old photograph from an album; it might seem like it transports you to the past, perhaps to your grandparents' wedding when they were both very young.

But the truth remains — you're viewing that picture in the present. Just as the watch's dial is a snapshot of the present, so are the moments and memories we hold dear, forever embedded in the fabric of the present.

In this contemplation of time, I've come to appreciate the enigma it presents.

But what is time?

Your Mickey Mouse watch measures the tension of a spring or the oscillations of a crystal, but it's not truly measuring time. Days are the rotations of the Earth, seasons mark our planet's orbit around our star, yet they, too, don't measure time.

In our existence, we find ourselves tethered to the present moment, the Now, representing the dynamic and non-deterministic realm of the universe. Let's embark on an exploration of the concept of the Now and its pivotal role in shaping our comprehension of time. As we delve into the emergence of the universe, the significance of time, the interplay between consciousness and information, and the limitations defining our existence, we'll continue to construct a narrative that may serve as a new perspective for this reality.

The Now is the active and non-deterministic region of the universe we call home.

The "physical" universe and all life only "physically" exist in the now.

Past and future frames of existence exist only as information characterized by how events (information) unfold and shape our perception of reality, with each moment being influenced by preceding moments (information).

Where we are now defines where we came from.

The events of the past, no matter how unlikely, are the route needed to arrive at this form at this moment. History, evolution, cosmic events, and the underlying rules of the universe are all emergent based on this Now.

The measurement of time is intertwined with the process of change and the presence of motion. We perceive time as past, present, and future, with our only access to it being the window of the present moment. Time allows us to observe and evaluate comparative changes, facilitating the existence and evolution of complex systems. While time is a real phenomenon to us as conscious observers, its exact mechanism remains elusive.

Time is a process of change and only has value to a conscious observer.

I believe our reality is composed of consciousness and information as its two primary dimensions. We perceive reality through a matrix of consciousness, where length, width, height, and time are emergent constructs built upon fractal information.

As we experience it, time exists only within the biological minds of conscious observers, with past memories, present observations, and future projections forming our understanding of time. Time is a process of change and only has value to a conscious observer; to us, time is a real phenomenon, a continuous change through which we live.

Time becomes evident through motion; sunrise, sunset, day into night, the changing of the seasons, and the movement of the celestial bodies all is indicative of continuous change.

An important aspect of time is the presence of motion, but also, time allows forces to act.

Imagine consciousness as the grand stage where a magnificent play unfolds. On this stage, you have information as the stage props and actors, and the script is time itself.

In this theatrical setting, think of information as the elements that populate the stage — the characters, objects, and scenery. They move and interact according to the script of time, playing out the story of our reality. The dimensions of length, width, and height represent the various aspects of the stage, creating a dynamic and ever-changing setting.

As conscious observers, we watch this captivating performance in the audience. The play unfolds before us, scenes transition, actors play their roles, and time progresses as the storyline. It's a continuous process, all orchestrated on the grand stage of consciousness, where information, like stage props and actors, enacts the script of time.

The aging process serves as a constant reminder of molecular change and the ongoing chemical interactions at play. Time provides the means for comparative evaluation and the existence and evolution of complex systems.

I believe our reality can be envisioned as a network of tiny fractal pockets containing information and governed by a simple set of rules. These pockets collectively interact, determining the possibilities and outcomes of the present moment.

Energy and matter exist as information within these pockets, with mass representing just one aspect. As the universe expands, there is pixelation and blurring, which influences the resolution of our reality.

Time is an integral part of our reality, intricately connected to the universe's expansion. It isn't a separate dimension but rather an emergent construct that arises from the transition from one present moment to the next.

As the universe expands, time moves forward.

The Now acts as the lens through which we perceive the unfolding of reality. It's a realm of perpetual change, where consciousness and information intertwine to shape our perception of time, space, and existence. Understanding the importance of the Now offers insights into the nature of consciousness, the emergence of universal constants, and the boundaries that define our reality.

As active participants in the collective consciousness, we shape the upcoming Now, perpetually evolving our comprehension and connection with the Universe.

Based on my observations and from my perspective, the fundamental constants of our reality emerged at the universe's inception and appear to remain unchanging and consistent over time. These constants are outcomes derived from a single force, and their exact values are meticulously adjusted to facilitate the existence of conscious life. This single force is most likely the one responsible for causing the expansion of our Universe.

The Anthropic viewpoint is a philosophical perspective that emphasizes the idea that the fundamental constants and conditions of the universe are finely tuned or adjusted to allow for the existence of conscious life, such as humans. It suggests that the universe's parameters are not random but appear to be precisely set to accommodate the emergence of intelligent beings capable of observing and understanding the cosmos.

Imagine our understanding of reality as a vast, intricate maze filled with twists and turns. In this maze, the fundamental constants are like the walls and pathways, defining the structure and limits of our journey. These constants, carefully arranged, create the maze we navigate.

"The universe is this field that generates limited conditions from infinite possibilities." — *Tiago Meurer*

Now, consider consciousness as the explorer within this maze. It moves through the pathways, seeking answers and insights. The constants are like the clues and markers guiding the explorer's path and experiences.

The Anthropic viewpoint acts as our compass in this logical maze. It points us in the direction of understanding, helping us navigate the complexities and mysteries of our reality. Just as a compass aids an explorer in finding their way, the Anthropic viewpoint guides us toward a deeper comprehension of the interconnectedness between these constants, consciousness, and the underlying workings of our Universe.

SIDE QUEST

Please see The Sorcerer's Paradox.

So far, so good, but it becomes tricky trying to explain the exact mechanism of the phenomenon of time. We, as individuals, perceive time as past, present, and future, with our only window to time in the present, "The Now."

In physics (similar to reality, hint hint), time plays a significant role in measuring motion and forces.

Could our universe be defined by just two dimensions: consciousness and information? Could it be that we live within a perceived matrix of reality, where length, width, height, and time are constructed as emergent properties of consciousness, utilizing fractal information?

Time is what we observe as the progression of events from the past through the present and into the future. Time is not a dimension. Time does not exist in the conventional sense; our experience of time only exists in the biological mind of a conscious observer, past as memory, the present as observation, and the future as extrapolated experience.

Causality operates within the transition from the present moment to the next, forming an integral part of the emergent framework that defines our reality. This framework itself arises from the expansion of the Universe.

Time, or the concept of change, serves as a means for progression and experiencing the present.

"When we listen to a hymn, the meaning of a sound is given by the ones that come before and after it. Music can occur only in time, but if we are always in the present moment, how is it possible to hear it? It is possible because our consciousness is based on memory and on anticipation. A hymn, a song, is in some way present in our minds in a unified form, held together by something ...by that which we take time to be. And hence this is what time is: it is entirely in the present, in our minds, as memory and as anticipation."
— *Augustine*

And just like gravity is a question, a clue. Time is information in motion.

Time is the progression of changed information in a processed format. Information from the past, present, and future (extrapolated data) is processed in a conscious observer's biological mind.

We give life to the enigma of Time.

We Are Defined by our Limitations
The many limits in our reality force a concentration of complexity and, in turn, gives us form.

We are shaped by our limitations, which compel us to navigate the intricate complexities of our reality and manifest our unique forms. These limitations encompass the dimensions we inhabit, the constant speed of light, the ever-flowing river of time, and the universal constants that govern the fundamental aspects of existence.

Now, let us delve into the essence of our being. I may suggest that we possess a dualistic nature, with our physical selves acting as vessels through which consciousness perceives the world. Simultaneously, our physicality allows consciousness to engage with and comprehend our reality, bestowing upon it reason and the freedom of choice.

Through the course of evolution, we have become the embodiment of this remarkable platform, where our personalities are a harmonious amalgamation of universal consciousness filtered through the imperfect lens of biology. While intelligence seems predominantly influenced by biological processes, the pursuit of beauty takes on a more ethereal, spiritual dimension.

Nevertheless, flawed biology can give rise to human weaknesses, perpetuated through erroneous reasoning and the establishment of man-made institutions founded upon flawed logic. Yet, as emergent constructs within the intricate tapestry of existence, we remain interconnected with a collective consciousness, actively shaping the unfolding of the present moment and co-creating the future. It is within the confines of our limitations that introspection is provoked, compelling us to question, contemplate, and ultimately strive for personal growth and improvement.

Paradoxically, because we must confront and live within our weaknesses, we find the inspiration and motivation to dream of our strengths. Acknowledging and accepting our limitations, we recognize the gaps in our understanding, skills, and capabilities.

These gaps can be viewed as spaces where the potential for growth and self-improvement resides, waiting to be filled with knowledge, experiences, and skills that enhance our strengths.
Our dreams emerge from envisioning a future where these gaps are bridged, where we have transcended our limitations and tapped into our fullest potential.

Our weaknesses become transformative opportunities in the realm of information, which lies at the core of our existence. Just as missing bits of information in a computational system can be replaced or filled, we can actively work towards filling the gaps in our own understanding and abilities. This process entails seeking new knowledge, learning from our experiences, and engaging in practices that foster personal development.

As we embark on this journey of self-improvement, our dreams expand and evolve. We begin to envision versions of ourselves that are more resilient, accomplished, and aligned with our inherent strengths. Living within our weaknesses becomes a catalyst for transformation, propelling us to dream bigger and reach higher.

Through the lens of the "Now," we recognize that our weaknesses represent spaces where new information can be integrated, leading to personal evolution and the fulfillment of our dreams.

Our existence isn't about reaching perfection but continuously improving.

The Conscious Anthropic Principle (CAP) provides further insights into the nature of our reality. It posits that consciousness is fundamental to the fabric of the universe and that the design of the cosmos is intentional, with the aim of nurturing intelligent life. The CAP invites us to explore the profound interplay between collective consciousness and the intricate web of fractal information, shaping the very foundations of existence. This perspective prompts us to contemplate the deliberate nature of the universe and the intrinsic interconnectedness between consciousness, the unfolding of the cosmos, and the emergence of life.

It may sound counterintuitive, but the purpose of our existence is not rooted in attaining perfection but rather in perpetual improvement.

Science is our tool for understanding how the world and universe work. It helps us describe, understand, and predict natural events. We gather evidence by observing and experimenting and we create strong models that can be reviewed and repeated. Through these models and thought experiments, we keep advancing our knowledge and exploring the unknown.

Newton's models were incredibly accurate and still hold true today. However, Einstein's special theory of relativity introduced some fascinating and revealing ideas about time. One of these ideas is the concept of "slowing of time" in the presence of a massive object or when observing motion. This means that the rate at which time passes can vary depending on your point of reference.

From your own perspective, time seems to pass at a constant rate but, if we compare the measurements of time by two different observers, we can see that time can actually move at different speeds. This phenomenon, known as time dilation, has been precisely confirmed by experiments.

The discovery that time can be slowed down is a groundbreaking clue that can lead us to a deeper understanding of reality. If we can comprehend why time dilation occurs, we may uncover the underlying nature of time itself. It opens up new possibilities for exploring the fundamental aspects of our existence.

The Twin Paradox is a fascinating thought experiment in the field of physics, specifically in the realm of special relativity. It involves a pair of identical twins, one of whom embarks on a journey into space aboard a high-speed spacecraft while the other twin remains on Earth. Upon returning home, the travelling twin discovers that the twin who stayed on Earth has aged more.

This paradox arises from Albert Einstein's theory of Special Relativity, which reveals that time is not experienced in the same way by everyone. To illustrate this concept, let's imagine that we have two clocks: our own personal clock and another clock that is moving at a constant speed relative to us.

In this scenario, the clock in motion, relative to ours, actually moves forward in time at a slower rate. This means that time progresses as the moving clock lags behind our own. The effect of time dilation becomes more pronounced as the speed of the moving clock approaches the speed of light, but even within our modern world, we can observe and measure this time difference.

In the context of the Twin Paradox, the twin who travels into space experiences time dilation due to the high-speed motion of the spacecraft. As a result, their personal clock runs slower compared to the clock of the twin who remains on Earth. When the travelling twin returns, they find that less time has passed for them than their Earth-bound counterpart, leading to the apparent age discrepancy.

This thought experiment highlights the intriguing nature of time and its relationship with motion. It demonstrates that the passage of time is not absolute but depends on the relative velocity of observers. Though a theoretical concept, the Twin Paradox helps us grasp the profound implications of Einstein's theory of Special Relativity and the intriguing nature of our Universe.

While the time dilation effect is significant only at extremely high speeds or in the presence of objects with high mass, its existence provides us with a deeper understanding of the fabric of reality and challenges our intuitive notions of time. Through experiments and observations, scientists have confirmed the predictions of Special Relativity, reaffirming the remarkable accuracy of Einstein's groundbreaking theory.

The Twin Paradox serves as a captivating illustration of the consequences of Special Relativity. It highlights the concept of time dilation, where the relative motion of observers affects the passage of time. This paradox showcases the fascinating and sometimes counterintuitive nature of our Universe, inviting us to delve deeper into the mysteries of space, time, and the fundamental laws that govern our reality.

We perceive change at a "sharp focal point" — the present (Right Now). As such, the past is just memory, and the future does not exist yet, which are mental constructs based on past experiences and observations. The passage of time and the feeling that past and future have some special existence, are illusions created by our mind trying to explain our world and are built by our ever-active ego.

I believe we exist in an evolving three-dimensional bubble of constructed consciousness. When we ask the question, *"What is time?"* we are really making inquiries that seek answers to the fundamental nature of reality itself, which leads us back to the question.

"Why are we here?"

Time is not a dimension. I believe, time is part of the emergent construct that is our reality and a product of the expansion of the Universe. The force that causes this expansion is the focal point of our reality, and all other forces are derivative products of this single force. The fractal heart of reality is emergent, expanding information. Conscious thought can only exist in the "Now," the "Past" can only exist as information, and the "Future" simply does not exist yet. This explains the arrow of time and why time travel is impossible.

We should really call the speed of light the speed at which information travels in our three-dimensional bubble reality. We are defined by our limitations. Time is necessary for progressive change and the evolution of complex forms. The speed limit of information is (not) the speed of light.

Information can travel in one of two ways, through the three-dimensional realm we live in or instantaneously via the dimension of information. This limitation is needed (paradoxically) to allow meaningful consciousness.

The idea that time might have a purpose and that this purpose could be connected to life is a thought-provoking concept. Some philosophical and scientific perspectives propose that time, in the context of our universe, could indeed be intimately linked to the emergence and evolution of life.

Let's explore:

Time as a Framework for Change: Time is fundamentally a framework for measuring change. Without time, there would be no sequence of events, and causality would lose its meaning. In this sense, time serves as a canvas upon which the story of life unfolds.

Life as a Product of Time: Life, as we know it, is believed to have evolved over billions of years. The passage of time allowed for the complex processes of biological evolution to take place, leading to the development of diverse and sophisticated life forms on Earth.

Consciousness and Meaning: If we consider the purpose of life, it might indeed be related to consciousness. Consciousness allows us to perceive, reflect upon, and find meaning in our experiences. We assign value, purpose, and significance to our existence through consciousness.

Purpose of Life as Meaningful Consciousness: From this perspective, the purpose of life could be seen as the cultivation of meaningful consciousness. It's about exploring the depths of our existence, understanding our place in the universe, and connecting with others profoundly. The pursuit of knowledge, wisdom, and the quest for meaning gives purpose to our lives.

Nonetheless, it's essential to recognize that the concept of purpose, particularly concerning the universe or time, remains inherently subjective and subject to diverse interpretations. Various belief systems, philosophies, and worldviews provide a spectrum of perspectives on life's purpose. What imparts meaning and purpose to an individual's life can vary significantly, reflecting the richness of human diversity. However, the very act of questioning the purpose of life is an inherent and self-evident facet of our shared human experience.

This odd perspective suggests a deep connection between time, life, and meaningful consciousness. It invites contemplation on the profound questions of existence and the role of time in shaping our journey toward understanding and purpose.

Time, the silent weaver of destiny's thread,
Unfurls its tapestry where life is led.
Its purpose, a dance with progress in embrace,
Nurturing growth, complexity's embrace.

Through its gentle sway, a symphony's rise,
A metamorphosis, where meaning lies.
From humble notes, a melody takes flight,
Birthing consciousness, profound and bright.

With every tick, a step on life's grand stage,
Time molds the clay as epochs turn the page.
In intricate forms, its magic weaves,
Unveiling wonders, where beauty retrieves.

Oh, temporal essence, elusive and wise,
Unveiling the secrets that rise within us.
For in your passage, our souls find release,
A canvas for creation, where wonders never cease.

At this point, I believe I can state:

- Successful life has more offspring.
- The emergent process of life is moving towards observation.
- Intelligence allows life to have a richer and fuller experience of consciousness.
- Human beings seem to have free will or choice.
- The universe we observe seems perfect for us, which implies an anthropic bias.
- The whimsical labyrinth of our reality is very complex and involves ideas and scientific understanding beyond mankind's current abilities.
- Reality has a predetermined structure rooted in the evolving fractal blueprint of information.
- That fractal blueprint of information can be shared.
- Time is a process of change and only has value to a conscious observer.
- Observation and intelligence are necessary for meaningful consciousness.
- Meaningful consciousness creates the structure (texture) of our reality from information that otherwise exists in an undetermined state of probability.

The Poetry of Eternal Time:
Our Role in the Unfolding Moment

In the grand tapestry of life, time is the invisible thread that weaves through every experience, from the ordinary to the extraordinary. It's a river that never stops flowing, carrying us along its currents. Yet, within this ceaseless motion, exists a profound stillness — the moment. This moment is where the poetry of eternal time unfolds, and we are both the authors and the readers of this timeless verse.

Imagine time as a never-ending book with infinite pages. Each page represents a moment, a slice of existence. These moments are strung together like pearls on a necklace, forming the story of our lives. We, as individuals, play a pivotal role in both writing and interpreting this story.

The present moment is where the magic of existence truly happens. It's a blank page waiting for us to inscribe our thoughts, actions, and emotions. This canvas of Now is where the past and future converge, creating a bridge between what has been and what could be. It's a canvas we are privileged to paint with our choices and intentions.

Consider the act of kindness — a simple gesture that can change the course of someone's day. In that brief moment of compassion, we influence the narrative of time. We become the authors of a verse filled with warmth, connection, and empathy. It's as if we've added a stroke of colour to the larger masterpiece of life.

Our lives are made up of countless such moments. Some are fleeting, like a smile exchanged with a stranger, while others are more enduring, like the bonds we form with loved ones. When infused with sincerity and presence, these moments become verses in the poetry of eternal time.

What's beautiful is that we have agency in this ongoing narrative. We're not merely passive readers, we are co-authors shaping our own stories. Every choice we make, every dream we pursue, and every connection we nurture becomes a line in our personal poem within the larger anthology of human existence.

As we journey through the river of time, let's remember that we are both poets and poems. We have the power to create verses that resonate with meaning, purpose, and love. The canvas of eternity is ours to paint, one moment at a time. Our lives, in their entirety, become a testament to the beauty of the poetry of eternal time — the intricate dance of moments, choices, and connections that make our existence a masterpiece in the grand tapestry of the Universe.

> *"Does time have a purpose, and could that purpose be life? If this is true, then the purpose of life is meaningful consciousness."* —— *Odd*

WHY ARE WE HERE?

Question 4

The ultimate question regarding the purpose of our existence loomed large, circling back to all the others in search of a resolution.

I've been pondering the idea of a "Book of Questions." I want it to be a theme that invites people to ask questions. It struck me that with everyone focused on writing books filled with answers, there should be a market for someone to craft a book centred entirely on questions. It's partly because of my sarcastic personality, I thought, "Yeah, that's me." I liked the concept of creating a book that encourages people to ask questions.

> *"I would rather have questions that can't be answered than answers that can't be questioned."* — Richard P. Feynman

Cold Beer on a Hot Day

There's a phrase that often comes to mind: "cold beer on a hot day." It's a simple, comforting image that captures the essence of joy on a summer afternoon. The contrast of the heat of the day and the relief of the cool beverage. In much the same way, I believe that our current reality is a tapestry woven from emergent events, each thread adding to the vibrant mosaic of the present moment.

Yet, if we peel back the layers, we uncover a deeper truth. Our reality, our experiences, are intricately intertwined with information, the threads that bind the fabric of existence. It's this interplay of events and data that crafts the narrative of what we perceive as the future — a realm that remains open to possibilities, much like a roll of the dice.

In this grand game of existence, I am convinced that the future is not rigidly predetermined, not etched in immutable stone. Under specific circumstances, it becomes the fertile soil for the growth of narratives, stories that could rival the most captivating of sci-fi novels. It's as if the cosmos itself is a master storyteller, weaving tales of stars, planets, love, bravery, and sentient beings into an ever-evolving saga.

I suggest, the universe is a thinking entity — a consciousness that contemplates its own vastness. Just as we ponder the mysteries of existence, the Universe, too, ponders its own nature. In this reflection, we find ourselves as integral components of this cosmic contemplation. We are not mere spectators, we are participants woven into the very fabric of thought.

I draw an analogy between our role in the universe and that refreshing cold beer on a scorching day. We might appear modest in the grand scope of the cosmos, like a single dewdrop in a vast ocean. People might whisper, "Perhaps that's all we are." However, I beg to differ. Our existence, thoughts, and actions are the ripples reverberating through the cosmic sea. We are the nuanced brushstrokes that paint colours onto the canvas of the Universe.

As we contemplate our place in this grand scheme, a realization dawns. Life, with all its complexities and intricacies, is a gift. Amidst the uncertainties and the vastness, we have the privilege of experiencing existence. Though it may sometimes seem humble, this privilege is something to be cherished and celebrated.

So, while we might ponder our cosmic role, let's not forget the significance of our individual stories. Like characters in an epic novel, we each contribute to unfolding this universal tale. The future is indeed up for grabs, and as we move forward, let us remember that we possess the power to co-author our destinies.

Just as the universe thinks, contemplates, and evolves, so do we. Like cold beer on a hot day, our lives offer moments of refreshment and delight, reminding us that despite the vastness of the cosmos, we are here, living, breathing, and experiencing — an opportunity that is, indeed, not so bad.

The Longing for Purpose: Exploring our Existence

Deep within us lies an inherent longing for purpose, a yearning to understand why we are here. This contemplation stems from our insatiable curiosity, driving us to delve into the profound mysteries at the core of our existence. From the celestial wonders of the night sky to the intricate balance of our planet, the universe seems to be meticulously crafted, igniting our quest for understanding. As we ponder the remarkable perfection of our surroundings, we begin to question the purpose behind our existence and contemplate whether our purpose is simply to be alive.

When we gaze at the stars above, we are captivated by their beauty and harmony. The intricate web of galaxies and the enchanting glow of stars reveal a universe that appears precisely crafted as if awaiting our presence. The delicate balance of natural forces, and the cosmic dance of celestial bodies all seem to align flawlessly. It is this awe-inspiring perfection that sparks our contemplation of purpose.

Let us consider the everyday miracles that surround us on Earth.

Our planet orbits the Sun at the perfect distance, allowing for the existence of life as we know it. The Sun itself, classified as a G-type main-sequence star or a yellow dwarf, provides us with the energy and warmth necessary for our survival, bathing us in a seemingly endless supply of low entropy. Moreover, our Moon stabilizes the Earth's movement, influencing tides and supporting the circulation of our oceans, which are essential for sustaining life.

> *"Entropy makes things fall, but life ingeniously rigs the game so that when they do they often fall into place."* — *John Tooby*

Earth possesses an unusual protective magnetic shield, shielding us from the harmful effects of solar radiation and charged particles. This shield plays a crucial role in maintaining the conditions necessary for life to thrive. Additionally, our planet boasts vast oceans that cover 71% of its surface, providing a habitat for diverse marine life and serving as a vital source of sustenance for all organisms.

The Earth's green landscape and complex ecosystems testify to the emergence of life and its continuous journey of observation. The emergent process of life (EPL) has led to the formation of intricate webs of life, enabling the interdependence and sustainability of various species. This symphony of life supports us, providing the air we breathe, the food we eat, and the energy we require for our existence.

When we contemplate the fundamental constants of the Universe, we observe an intriguing phenomenon. These constants, such as the speed of light, Planck's constant, and the gravitational constant, appear precisely tuned to allow for the emergence of life. This observation, known as the anthropic principle, hints at a purposeful alignment of the universe to facilitate our existence.

In our quest to understand our purpose, we find ourselves immersed in a universe that is both elegantly simple and astoundingly complex. The harmonious alignment of celestial bodies, the intricate balance of natural forces, and the many nurturing aspects of our planet all point towards a purposeful existence. While the question of our ultimate purpose may remain open-ended, we can appreciate the beauty and awe-inspiring nature of our surroundings. Perhaps, in the end, our purpose lies not only in pondering the mysteries of the universe but also in appreciating the remarkable gift of life itself.

"Knowledge is the foundation of our existence. Love is its purpose."
— *Odd*

WE SEE FROM THE INSIDE OUT

Picture yourself awakening on a Saturday morning, as the initial rays of light seep through the blinds.

What are you experiencing?

Normal circadian rhythm, in your nice warm bed?

Or, flying through the cosmos on a small iron, nickel planet, spinning at roughly 1,609 kilometres per hour, around a small even-tempered, yellow dwarf star, at 107,826 kilometres per hour, on the outskirts of a spiral galaxy spinning 209 kilometres per second, in a bubble universe which appears to be expanding faster than the speed of light?

Or are you still dreaming?

Everything must have some value defined to be perceived and that requires a conscious observer. Observation is surprisingly subjective and is orchestrated by an enormous number of variables. The human mind is a vast array of billions of neurons, linked in a web-like network of regions and areas of specialization. The brain takes in information on a scale difficult to describe and any technical attempt to summarize this flow will fail to convey this everyday wonder. This information is routed, processed, stored, and rerouted millions of times every second. The "Mind" seems to have a hierarchy of activities

The human brain, when viewed from a top-down cross-section, can be divided into three major components that align with what is known as the golden circle. Our newest brain, the neocortex, corresponds to the "what" level. It is responsible for rational thinking, analytical thought, and language abilities. The neocortex enables us to reason, analyze information, and communicate through words.

The two middle sections of the brain make up our limbic brains. These limbic brains are responsible for our emotions, such as trust and loyalty. They also play a crucial role in human behaviour and decision-making processes. Unlike the neocortex, the limbic brain lacks the capacity for language. This means that when we communicate with others, decisions are actually happening in the limbic brain, which controls decision-making but not language.

In simpler terms, the human brain can be thought of as having three main parts: the neocortex, which handles rational thinking and language, and the limbic brains, which deal with emotions, behaviour, and decision-making. While the neocortex allows us to think logically and express ourselves through words, the limbic brain guides our decision-making processes, even though it cannot use language for communication.

Understanding these different components of the brain helps us appreciate the complex workings of our minds. It highlights the interplay between rational thought and emotional responses, shedding light on why we make certain decisions and how our brain functions in relation to language and behaviour.

Conscious man, with his complex mind and brain, is the remarkable outcome of billions of years of purposeful effort of the emergent process of life and the tireless effort of life to improve. Using DNA and the hammer of survival, life has shaped all forms of life on this planet. Our existence is built upon the simple building blocks inherited from our primitive ancestors.

Throughout the evolutionary journey, life has relentlessly strived to improve and adapt. Genetic variations have been tested, and those best suited for survival have been passed down through generations. Our conscious experience, with its profound complexity, has emerged as a result of this ongoing process.

The mind, encompassing both conscious and subconscious aspects, exerts comprehensive control over our actions and perceptions. The brain, responsible for regulating mundane bodily functions like heart rhythm and facilitating extraordinary capabilities such as creativity, plays a crucial role in shaping our experiences. Moreover, the mind constructs our perception of reality, with a key element in this process being the ego.

The ego, a hypothetical construct, serves as the intermediary between the conscious and unconscious realms of the mind. It integrates available information and fabricates our overall view of reality.

The term "ego" originated from the Latin word for "I" and gained prominence through the work of Sigmund Freud, even though he himself used the German term "das Ich," meaning "the I."

In psychology, the ego is further dissected into three components: id, ego, and superego.

The id represents the primitive and instinctive aspects of personality, comprising inherited and biological elements present from birth. It operates in an irrational and chaotic manner, driven solely by desires and pleasure-seeking impulses. To mediate between the id's unrealistic demands and the external world, the ego develops later and serves as the decision-making component of personality. Unlike the id, the ego functions by reason, seeking realistic ways to satisfy the id's desires while considering the constraints imposed by reality and society.

The superego emerges during childhood development, typically around the ages of three to five years. It incorporates the values, morals, and societal standards acquired from the individual's environment. The superego acts as a moral compass, rewarding behaviours aligned with these standards with feelings of pride and satisfaction while punishing deviations with feelings of shame and guilt. The specific response depends on which part of the ego (the ego ideal or conscious) is activated.

Operating in accordance with the reality principle, the ego navigates the complexities of satisfying the id's demands while minimizing negative consequences imposed by society. It takes into account social realities, norms, etiquette, and rules when determining how to behave.

The mind, encompassing conscious and subconscious aspects, exerts comprehensive control over our actions and perceptions. The ego, as the intermediary between the conscious and unconscious realms, plays a pivotal role in shaping our view of the world. The id represents the primitive and instinctive elements, the ego functions as the decision-making component, and the superego incorporates societal values and morals. Understanding these components provides insights into how our minds operate, enabling us to navigate social dynamics and make choices that align with both our individual desires and the expectations of society.

> *"Reality is just a conspiracy theory devised by the imagination to keep us from questioning the absurdity of existence."* — Salvador Dali

The human ego is the stage in which I believe we view consciousness. I also believe supporting this stage is our primary purpose.

The human ego, in its essence, serves as the stage upon which consciousness unfolds. It is a remarkable platform that enables us to navigate the complexities of life. I firmly believe that our primary purpose lies in not just supporting but enhancing this stage for the betterment of ourselves and society.

Imagine the human ego as the spotlight in a grand theatrical production. It's where we all take centre stage, each with our unique roles and stories to tell. Our thoughts, emotions, and actions are the scripts, and how we perform on this stage can shape the quality of our lives and the world around us.

Promoting better "acting" on this stage means encouraging positive, constructive behaviours. It's about fostering self-awareness to recognize when our ego may lead us astray, causing unnecessary conflicts or misunderstandings. By taking a more mindful approach to our actions and interactions, we can improve the overall quality of our performance.

Moreover, this purpose involves extending kindness and understanding to others as fellow actors in this grand production. When we offer support and encouragement, we create a more harmonious and cooperative atmosphere, much like a well-rehearsed ensemble cast working in unison.

In this grand play of life, our primary role is to uplift one another and to create a performance filled with compassion, empathy, and cooperation. By doing so, we not only enhance the human ego's stage but also contribute to a more positive and harmonious world for all to enjoy.

The mind is also home to our sense of personal identity. *What makes up your personality, and which personality traits do you have?* Personality traits can be thought of as individual differences in behaviour, thought, and emotion. Personality may be part of our unconscious. It may consist of personal narratives that we build across our lives, or it may be in part, the observable manifestation of our genetics.

Regardless of where personality comes from, I believe it can be helpful to be aware of our own traits and that understanding may help us understand others.

In the tapestry of human existence, we see from the inside out. Our consciousness emerges from the subjective realm of perception, influenced by the interplay between the mind and ego. Understanding the intricacies of our personal identity and its impact on our interactions with others is key to fostering empathy and compassion.

As we awaken each day, we possess the power to expand our awareness beyond the ordinary, embracing the extraordinary complexity within ourselves and the world around us.

> *"All the world's a stage."* — *William Shakespeare*

ODD FACT

Dolphins and apes have a much wider range of genetic differences among their groups compared to modern humans.

Think of chromosomes as the chapters in the book of life. They hold the special instructions that make every living thing one-of-a-kind. But here's the cool part: different creatures have different numbers in these chapters. Dolphins have 44 chapters, apes have 48, and humans? We've only got 23.

Picture life's story as a giant book where each chapter (or chromosome) plays a crucial role. In our book, humans have fewer chapters than dolphins or apes. But don't let that fool you, it doesn't mean our story is any less exciting — it's just a different kind of story.

Now, let's talk about our ancient family. A long time ago, about 900,000 years back, our early human ancestors faced some tough times. They had to deal with things like droughts and climate changes. These challenges made their population super small — so small that there were only about 1,280 adults who could have kids.

This was a bit like a cliffhanger in the story of human life. Scientists call it a "population bottleneck." It's like our story almost reached its end, with only a few characters left.

But here's the crucial point: all modern humans are incredibly closely related due to this challenging time in our history. We share a common ancestry, and when we look at our genetic differences, they are relatively small compared to some other species.

Scientists have used their fancy tools to explore our family tree and figure out what happened back then. They discovered that this population crash might have led to new branches in our family tree. It's almost as if the story took an unexpected turn, and new chapters started to form.

So, in simpler terms, think of chromosomes as chapters in the book of life, and each species has its own set of chapters. Our early human ancestors faced a tough time when their population got incredibly small, almost like a "to be continued" moment in our story. But thanks to science, we now know that this challenging period might have added some surprising twists to our family tree, making our story even more fascinating. And remember, despite our differences, all modern humans are part of one big family.

I hold a strong belief in the idea that we are all but one facet of the gem of light that is collective consciousness. If this belief is correct, it should inspire us to treat each other as cherished siblings.

The notion that we are all part of the same collective consciousness is profoundly moving. It's as if the very fabric of existence weaves us together into a tapestry of shared experience. In light of this interconnectedness, we should embrace the idea of treating each other as siblings. In a family, our differences become threads that add diversity and richness to the tapestry of life. Just as siblings may have different personalities and interests, so do we as human beings. But despite these differences, there's an underlying bond, a shared heritage, and a sense of belonging.

Imagine a vast cosmic family, where every individual is a sibling. This concept evokes a deep sense of empathy and compassion. When we view each other as siblings in the grand family of humanity, it becomes natural to care for one another. We become more inclined to understand each other's struggles, celebrate each other's successes, and lend a helping hand when needed.

It also brings a sense of responsibility. Just as siblings look out for each other, we should strive to create a world where no one is left behind. This shared responsibility encourages us to work collectively for the betterment of all, addressing issues like inequality, injustice, and environmental concerns.

Embracing the idea of treating each other as siblings is a call to foster a sense of unity and togetherness. It's a recognition that our individual well-being is intricately tied to the well-being of others. By extending kindness, respect, and love to our cosmic siblings, we not only enrich our own lives but also contribute to a more harmonious and compassionate world where all can thrive.

Right Now

The Universe, with its intricate tapestry of existence, is a profound manifestation of consciousness, revealing itself as an emergent form of information. As we embark on this journey, let this guide serve not only as a manual but as a companion, offering a new perspective for this reality and providing guidance to unravel the mysteries that shape our anthropic experience.

Our purpose, rooted in perpetual improvement, seeks to transcend our biological limitations and embrace the spiritual dimensions of our existence. By fostering knowledge, love, and the pursuit of joy, we aim to create a positive impact on our surroundings and contribute to the collective consciousness of humanity.

Time and change provide the canvas for meaningful transformation. By being present and consciously shaping our reality, we unlock our potential for positive change. Promoting education, wisdom, and understanding enriches our collective consciousness and empowers us to enhance our purpose.

Finally, in the grand scheme of things, our fulfillment and joy hold greater significance than mere notions of winning, as they contribute to the fabric of existence and leave a lasting mark on the tapestry of consciousness.

Now, let's explore these steps in more detail to help start your journey of questions.

1. Perpetual improvement.

> *"Fate is like a strange, unpopular restaurant filled with odd little waiters who bring you things you never asked for and don't always like."*
> *— Lemony Snicket*

Do our numerous limitations shape our existence, giving rise to the complexity and form we perceive? Do the dimensions we inhabit, the speed of light, our relationship with time, and the constants of the universe establish the boundaries within which our reality unfolds?

Is it because of our biological imperfections, rather than in spite of them, that we possess remarkable abilities like learning, appreciating beauty, showing compassion, and love?

It may seem counterintuitive but I believe the purpose of our existence is not rooted in attaining perfection but, instead, in perpetual improvement.

These aspects transcend our mere biology and hint at the spiritual dimensions of our existence.

2. The desired outcome selected is the outcome that has a future and has the most net fulfillment for our collective consciousness.

"Knowledge is the foundation of our existence. Love is its purpose."
— *Odd*

The pursuit of happiness and the desire to make a positive impact on the world around us are fundamental aspects of human existence. The belief that we are meant to experience joy and have a net positive effect on our surroundings is derived from the interconnectedness between consciousness, time, personal growth, and the human ego. By understanding these concepts, we can uncover our purpose and strive for fulfillment and joy.

At the core of our existence lies the pursuit of joy. It is a universal human desire to find happiness, contentment, and satisfaction. By embracing joy in our lives, we not only fulfill our own needs but also create a ripple effect that spreads positivity to those around us. Joy is a powerful force that uplifts spirits, enhances relationships, and cultivates a harmonious environment.

The desired outcome of each cycle would be joy and enlightenment. Pain and suffering are not permanent; in fact, nothing is. This may seem paradoxical, but, like how evolution is not truly the survival of the fittest but rather the ability to spread one's genes to future generations, the selected desired outcome is the one that holds a future and offers the most net fulfillment for our collective consciousness.

3. To create time and allow change.

> *"Does time have a purpose, and could that purpose be life? If this is true, then the purpose of life is meaningful consciousness."* — Odd

To create time and allow change. Time is a precious resource that allows for growth, transformation, and the unfolding of possibilities. As we navigate the river of time, we have the opportunity to create meaningful change in ourselves and the world. Change is inherent in our journey, and it provides avenues for progress, innovation, and the realization of our potential. By recognizing the significance of time and embracing change, we contribute to the overall evolution of consciousness. An outcome without a usable conscious platform would be avoided.

4. To be aware of the present and create positive change.

> *"Information and consciousness form the very framework of our reality. We awaken each day to view the world open in front of us unaware how accurate that metaphor is."* — Odd

Do the emergent characteristics of history, evolution, cosmic phenomena, and the underlying principles of the universe revolve around the present moment? Could the present moment be an extraordinary factor that steers the course of both the past and future? The awareness of the present and its capacity to foster positive change is a central theme. Consciousness, our subjective experience of being aware, plays a crucial role in defining our purpose. It is through consciousness that we become aware of the present moment and our ability to influence it. By cultivating mindfulness and gaining an understanding of our thoughts, emotions, and actions, we can consciously choose to create positive change. Each decision we make, no matter how small, holds the potential to shape our reality and the world around us.

5. To promote improvement.

> *"We live on an island surrounded by a sea of ignorance. As our island of knowledge grows, so does the shore of our ignorance."*
> — *John Wheeler*

Promoting education, wisdom, and understanding is crucial for fulfilling our purpose. Learning expands our knowledge, broadens perspectives, and deepens self-awareness. It empowers us to pursue joy, embrace change, and make informed decisions. Sharing wisdom and fostering a culture of continuous learning enriches the collective consciousness of humanity. We can use this insight to enhance the first four.

6. Actively shape our reality.

> *"The illusion that the self and the world are broken into fragments originates in the kind of thought that goes beyond its proper measure and confuses its own product with the same independent reality. To end this illusion requires insight, not only into the world as a whole, but also into how the instrument of thought is working. Such insight implies an original and creative act of perception into all aspects of life, mental and physical, both through the senses and through the mind, and this is perhaps the true meaning of meditation."*
> — *David Bohm*

The human ego unfolds as a captivating stage where consciousness takes centre, offering a unique platform to navigate life's intricacies. Our primary purpose is to support and enrich this stage for personal and communal betterment.

Picture the human ego as a focal point, much like a warm spotlight illuminating our individual roles and stories. Our thoughts, emotions, and actions are the scripts, weaving the fabric of our lives and influencing the world around us. Embracing positive behaviours, nurturing self-awareness, and refining our performance contribute to this purpose.

As we extend kindness and understanding to others, recognizing them as fellow actors, we create a harmonious and cooperative atmosphere, akin to a tightly-knit ensemble.

In this intricate play of life, our primary role is to uplift one another, creating a heartwarming performance filled with compassion, empathy, cooperation, joy, and love. Through these shared moments, we enrich the human ego's stage and contribute to a more positive and harmonious world for all.
Universal Consciousness and Shared Fractal Information are not just abstract concepts, they are like beacons that guide us through the enigmatic fabric of existence, guiding us with expanding questions.

The human ego is often perceived as the platform on which consciousness manifests itself. Although the ego is sometimes negatively regarded, it plays a vital role in fulfilling our purpose. It serves as a means for self-awareness and personal growth, enabling us to comprehend our own selves, desires, and our interconnectedness with the world. By nurturing and supporting the development of the ego, we establish a solid groundwork for self-discovery, empathy, and compassion.

Recognizing the profound link between the inner workings of the universe and the intricate functions of the brain expands our understanding of how the universe's emergent properties may have influenced the evolution and capabilities of the most intricate structure known to us. Moreover, it raises a fascinating possibility, the relationship between physics and cognition might be mutually influential.

As we delve into how the rules of the universe correlate with brain function, we gain insights into the purposeful design and interdependence between the fabric of the cosmos and the workings of our minds. This realization offers glimpses into the connection between physics and cognition, shedding light on the reciprocal influence between the functioning of our minds and the underlying principles governing the Universe.

The universe is fundamentally grounded in mathematics and can be perceived as an emergent form of information. The human disciplines of physics, mathematics, numbers, golden ratios, and even our constructed models and principles appear to describe our Universe.

Is this a mere coincidence? Or do our minds actively shape our reality?

7. No winning, only finding fulfilment and joy.

> *"What you think, you become.*
> *What you feel, you attract.*
> *What you imagine, you create."*
> — *Buddha*

The Fulfilment of Purpose: In the grand scheme of existence, there is no ultimate winning or losing. Instead, our purpose lies in finding fulfilment and joy. We contribute to the collective consciousness and the overall well-being of humanity through the pursuit of joy and the positive impact we create. By aligning our actions with the principles of joy, growth, consciousness, and wisdom, we find purpose and make our mark on the tapestry of existence. There is no winning, only finding fulfilment and joy.

Right Now

In the vast expanse of existence, a singular moment stands out, a moment that encompasses the totality of our reality. This moment is "Right Now." It is in the embrace of "Right Now" that we find ourselves entangled in the profound mysteries of our existence. Let us embark on a journey of exploration guided by this enigmatic moment.

In contemplating the notion of self-emergence, we envision reality as a magnificent tapestry, with each thread symbolizing a piece of information. We are both the weavers and threads of this ongoing work of art. Could it be that our reality spontaneously emerges, akin to a masterpiece weaving itself into existence?

Within this contemplation, consciousness assumes centre stage. It is not a product of our cognitive abilities but serves as the backdrop of everything. We are a part of consciousness, not the other way around. Our role is that of observers, intricately woven into the fabric of consciousness. It has always existed, and we bear witness to its unfolding.

As we navigate through "Right Now," we delve deeper into our role in the grand cosmic narrative. *Are we the avatars of consciousness and manifestations of information?* The idea of emergence continues to captivate us. Everything seems to emerge, with consciousness casting its profound reflection on it all. We are not the architects of consciousness but rather participants in a cosmic play. Consciousness predates our existence, and our purpose lies in observing and experiencing its intricate dance.

Time, a concept that weaves itself seamlessly into our existence. Right Now, we grapple with the enigma of time. It is not merely a passive element in our lives; we are both its creators and carriers. Time finds its dwelling within us, and we share it with the universe. Our function involves processing information, and through this process, we gain the ability to experience time. We become the sands in the hourglass, each of us a small but integral part of the vast tapestry of existence.

In the ever-evolving narrative of "Right Now," we contemplate the concept of an emergent event. Our consciousness, a complex tapestry of information, plays a pivotal role. As our intricate brains navigate through the realms of consciousness, they generate the essence of time itself. In this intricate dance, we experience everything one frame at a time, each "Now" unfolding in perfect synchrony. It is in the embrace of this singular moment, "Right Now," that we find our purpose. Nothing else exists in this eternal "Now" that is the heartbeat of existence itself.

"Right Now" is more than just a moment in time. It is the portal to the mysteries of our existence, a profound tapestry woven from the threads of self-emergence, consciousness, time, and purpose. As we traverse the depths of "Right Now," we unravel the enigmatic truths that lie at the core of our reality, forever guided by the eternal pulse of the present moment.

> *"The purpose of life is the grand defiance of the inevitable.*
> *The end is not the journey; it's just the last step.*
> *The journey is up to you to decide." — Odd*

Our lifetime is a precious, unrepeatable work of art.

The awareness that we seem to have just this one lifetime in our current form is both poignant and humbling. It's a singular journey, a once-in-an-eternity experience that we get to live through. This realization stirs a mix of emotions — from wonder and awe to a subtle tinge of melancholy.

Each lifetime is like a precious, unrepeatable work of art. While certain patterns and rhythms in life might appear cyclical, the specific brushstrokes and colours that create our personal narrative are entirely unique. It's as if we are all the authors of our own novels, with each page filled with stories, experiences, and emotions that will never be replicated.

This awareness of life's singularity is a reminder of the preciousness of the present moment. It encourages us to savour every experience, to treasure the fleeting moments of joy, and to learn from the challenges that shape us. It pushes us to make the most of the opportunities that come our way, knowing that once this lifetime is over, it won't return in the same form.

The idea of time travel, while a tantalizing concept in science fiction, remains beyond the boundaries of our human experience. Yet, it underscores the importance of the here and now. We cannot go back and change the past, nor can we leap into the future to see what's in store. What we have is this singular journey through time, and it's up to us to make it as rich, meaningful, and fulfilling as possible.

"If you are lucky enough to find a way of life you love, you have to find the courage to live it." — John Irving

In the grand tapestry of existence, our lives are the threads that weave together, creating a mosaic of experiences, emotions, and stories. Embracing the uniqueness of our individual lifetimes, we can find a profound sense of purpose and appreciation for the beauty that surrounds us. It's a reminder to live with intention, love wholeheartedly, and leave a positive mark on the world, for this journey is a one-time gift that should be cherished with all the depth of our human emotions.

SIDE QUEST: THE STORY OF MARY

Once upon a time, in a distant kingdom, there lived a beautiful and brilliant princess named Mary. But alas, she found herself trapped in a tower, far away from the vibrant world outside. This tower was a gloomy place where only shades of black and white were allowed. Her captors were eccentric scholars, studying the mysteries of perception.

Mary had spent her entire life locked away in that tower, devoid of colours. However, her thirst for knowledge was insatiable. She devoured books, conducted experiments, and engaged in deep discussions about the wonders of colours. She knew all there was to know about the physics, biology, and psychology of colours. But there was one thing she had never experienced herself — she had never actually seen colours.

One fateful day, the captors decided to reveal a long-kept secret to Mary. They led her to a door she had never been allowed to enter before. As the door swung open, Mary's eyes widened with awe and wonder. Before her lay a room bathed in the most enchanting shade of red. It was as if the room itself was alive, pulsating with warmth and energy, casting a magical spell on anyone who beheld it.

As Mary stepped into the room, a surge of emotions coursed through her veins. The colour red enveloped her senses, igniting a flame of passion within her. It was a revelation — a glimpse into a world she had only read about but never truly understood. The experience awakened something deep within her — a realm of existence that surpassed mere knowledge.

In that moment, Mary discovered the concept of qualia — the personal and subjective feelings that arise from our perceptions. She realized that her vast scientific knowledge about colours paled in comparison to the actual experience of seeing them. It was not just about facts and figures, it was about the indescribable qualia — the unique sensations and emotions evoked by the colours she beheld.

The red room became a symbol of Mary's newfound understanding. It opened her eyes to the wonders of subjective experience, reminding her that there were aspects of life that could not be fully explained by science alone. It was like trying to capture the essence of a delectable feast or the warmth of a loving embrace — those were moments that could only be truly grasped through personal encounter.

From that day forward, Mary's perception of the world transformed. She embraced the power and beauty of qualia, those magical feelings that added depth and richness to our existence. Her story became a beacon of inspiration, reminding others of the extraordinary capacity within each of us to embrace the mysteries and marvels of our own subjective encounters.

And so, Princess Mary ventured out of the red room, carrying with her a newfound appreciation for the complexities of perception. She embarked on a journey through the kingdom, marvelling at the kaleidoscope of colours that adorned the world. With each new hue, she cherished the unique qualia it brought forth — those unexplainable, breathtaking sensations that made life truly extraordinary.

As she shared her tale with others, they, too, were inspired to seek out their own moments of discovery. Together, they set out on a quest to uncover the enchantment of qualia that lay waiting to be explored in the vast tapestry of human experience. And in their shared exploration, they found a deeper connection to the world around them, embracing the magic that colours and qualia brought to their lives.

Our Embrace of Infinity

Light of Consciousness ↔ Fractal Information (Both Fundamental)
↓
$E = mc^2$ (Everything is Energy), (Secondary)
↓
Energy → Transforming Fractal Information
↓
Emergence → (The Guiding Hand that Organizes Fractal Information and Forms the Intricate Fabric of Our Reality)
↓

Time → The Driving Force of Change
↓ (Through the Lens of a Conscious Observer)
Right Now ↔ Information Preserved in the Present Moment
↓
Participatory ↔ Evolution (EPL)
↓
(Insert Infinity Here)
(Just making sure you were paying attention)

The light of consciousness giving form to fractal information suggests that everything in our reality possesses inherent energy and operates within a structured system. Energy, as a transformative agent, shapes and organizes fractal information, guided by the hand of emergence. Time, when viewed through the eyes of a conscious observer, represents the driving force behind change and the increasing complexity of reality. It is through consciousness that perception occurs, assigning meaning and significance to information. This intricate interplay between consciousness, evolving information, energy, and time weaves the complex tapestry that is our reality. The phrase "our embrace of infinity" signifies our acknowledgment and acceptance of the limitless nature of existence, where boundless possibilities and potentials unfold within the interconnected realms of energy, information, and consciousness.

Information is fundamental, reality is just the details.

Welcome to the emergent event of Right Now.

Mona Lisa's Smile

Imagine a sweeping panorama of existence, where universal consciousness acts as the expansive canvas. Fractal information dances across this canvas, serving as both the vibrant colours and the precise brushstrokes that intricately form the composition. Time takes on the role of the dedicated artist, with each passing moment being a deliberate stroke that adds depth and richness.

Within this dynamic artwork, emergent reality finds its place through humanity's presence. Like posing models, we contribute to the evolving narrative, bringing our unique experiences and perspectives to the forefront. We are the embodiment of Mona Lisa's smile — a focal point that captures the essence of the human experience.

SIDE QUEST

Colours of objects, usually viewed when they are illuminated by white light, such as sunlight or ordinary room light. White light is a mixture of all colours, roughly in equal proportions. White objects appear white because they reflect all the visible wavelengths of light that shine on them, resulting in the light still appearing white to us.

On the other hand, coloured objects reflect only specific wavelengths and absorb the rest. For instance, when white light shines on a red ball, the ball reflects mostly red light, causing us to see the colour red. The ball absorbs most of the greens and blues present in white light, making them invisible to us. Similarly, a blue ball reflects the blue part of the white light spectrum, while absorbing the red and green parts.

What happens when a red light shines on a red ball? The ball continues to reflect the red light, so it remains red. However, a white ball would also appear red in red light because it reflects all colours. If we shine blue light on a red ball, it will appear dark because it does not reflect blue light. It cannot appear red unless there is red light coming from the light source, and it cannot appear blue because the red ball absorbs blue light. Therefore, when we ask about the colour of an object, the answer is not simple, it depends on the colour of the light we use to observe it.

SORCERER'S SANDBOX

Welcome to the Sorcerer's Sandbox, a realm of curiosity and wonder where we embark on an enchanting journey guided by the allure of intriguing questions. In this whimsical playground, our imaginations roam free, leaving the mundane behind in favour of the extraordinary.

Here, we ponder the very nature of reality and the emergence of fundamental properties shaping our world. We explore the captivating notion that we might be nothing more than reflections of universal consciousness, constructed from the ever-evolving building blocks of fractal information captured in the moment. We contemplate our unique relationship with time — do we permit its existence, or does time reside within us? Ultimately, we arrive at the profound realization that "Right Now" is not merely a moment, it is everything. The amazing unfolding of reality, page by page.

As we journey through the twilight of our understanding, you'll encounter thought experiments and paradoxes that challenge your perspective on the world. Each attraction is designed to ignite your curiosity and guide you through the uncharted realms of existence. So, let's embark on this enchanting adventure into the heart of reality's mysteries and unveil the secrets concealed within the Sorcerer's Sandbox.

THE SORCERER'S PARADOX

"No wonder you're late. Why, this watch is exactly two days slow."
— Mad Hatter

The Sorcerer's Paradox unfurls a captivating narrative that transports us to a fantastical realm where a sorcerer wields the extraordinary power to cast a wish spell — an omnipotent force capable of reshaping the very fabric of the past. Rooted in the realms of fantasy, this paradox takes us on a profound exploration of questions about perception, reality, and the nuanced influence of observation.

At its core, the enchanting moment of invoking the wish spell initiates a transformative shift in the world, giving rise to a new reality. The intrigue deepens as this altered reality seamlessly integrates into the existing timeline, erasing any trace of the sorcerer's intervention. From the perspective of those immersed in this revised reality, it has always been this way. This prompts a compelling question: can anyone discern that a wish was made, given that the altered reality erases all memory or evidence of its own creation?

This paradox invites contemplation of the intricate dynamics between the act of wishing, the alteration of events, and the seamless integration of these changes into the timeline. It challenges our understanding of cause and effect, questioning the very nature of reality perception. If the past is reshaped by a wish in the present, leaving no discernible trace of its original form, does it create a reality indistinguishable from the one that preceded it?

This line of inquiry leads us to question whether an internal observer can genuinely discern changes in the fabric of reality if they occurred in the past. If the rules governing our reality underwent a cosmic U-turn, could we even notice? The paradox propels us to explore the boundaries of comprehension and the mysteries beyond the imagined limits of existence.

The Sorcerer's Paradox serves as an analogy: If historical information is stored in the emergent moment of now, can we discern changes in the past? Counterintuitively, I suggest that our perception may hold the key to understanding the intricate nature of our reality — the only reality that truly matters. Anything outside our observable history, to us, simply does not exist and remains confined to the realm of imagination.

In a universe constrained by the limits of information captured within the present moment, the exploration of something existing beyond those confines becomes a non-issue, beyond knowing, because it does not exist. The universe, inherently informational, springs to life through the act of observation — a notion resonating with interpretations in quantum mechanics and information theory.

As the paradox unfolds, it prompts us to ponder mysteries beyond the conventional confines of time and existence. The suggestion that our perception may hold the key to unlocking the intricate tapestry of our existence implies that nothing is truly unfalsifiable. In this context, past information, no longer part of our present, seemingly ceases to exist, tidying up reality in a very real way.

Extending this line of thought, the narrative ventures into a hypothetical scenario where the rules governing our universe changed in the past. The realization is that these changes would lack substance if not captured in the present moment. The past events, however improbable, are the necessary route leading to the present form we inhabit today.

Embarking on this intellectual journey prompts the recognition that our understanding of reality is an evolving exploration. The paradox challenges us to broaden our cognitive horizons and question the very essence of our existence. It invites us to delve into the intricate interplay between observation, perception, and the fabric of reality, pushing the boundaries of our philosophical contemplation.

In essence, the paradox encourages a bold blend of curiosity and ingenuity in the face of seemingly insurmountable unknowns that envelop us. It urges us to gain understanding and push past our imagined limits of comprehension, fostering a sense of curiosity that propels us toward the frontiers of theoretical physics and metaphysical inquiry.

Within this ongoing dialogue, profound insights may emerge, reshaping our understanding of the universe and, perhaps, influencing the very nature of our reality in the process. The enigmatic narrative of "Right Now" may be fluid and ever-evolving. I posit that if we can observe it, we can understand it; if we cannot observe it, it does not truly exist. In essence, information is primary, while reality merely consists of the details.

THE LIBRARY CONUNDRUM

"You can't have everything... where would you put it?"
— *Steven Alexander Wright*

In the whimsical world of cosmic contemplation, we are faced with a conundrum akin to the age-old question: If everything is information, where do you put all the data?

We imagine a cosmic library. This library houses the "Book of the Day," a tome that chronicles the events of the observable universe for each passing day. While the concept of this cosmic library is intriguing, it beckons us to ponder the profound challenges posed by the staggering volume of information it contains. From the movements of celestial bodies to the births and deaths of beings, from the dance of atoms to the journey of particles of light, every nuance of existence is inscribed within its pages.

To complicate matters further, consider that this library has been open for 14 billion years or that in a multiverse, every choice spawns new universes with their own unique libraries. Moreover, what if conscious individuals each possessed their personal "Book of the Day?"

The Vastness of the Cosmic Library

The notion of a cosmic library, with each day's events chronicled in a monumental tome, serves as a metaphor for the boundless expanse of the universe and the sheer depth of knowledge it holds. The idea that a single day's worth of universal happenings could fill a book of such magnitude is awe-inspiring. It invites us to contemplate the cosmic grandeur and the intricacies that unfold within it.

The Cosmic Timescale

To fully grasp the magnitude of the Library Conundrum, we must consider the staggering timescale involved. Our universe has been evolving for approximately 14 billion years, and each passing day adds another layer to the cosmic narrative. This temporal perspective accentuates the overwhelming nature of the library's contents. It urges us to reflect upon the ceaseless flow of time and the relentless accumulation of knowledge.

The Multiverse Complexity

The multiverse theory further complicates the Library Conundrum. According to this theory, each choice spawns new universes, each with its distinct library of events. This implies the existence of an infinite number of libraries, each documenting a different facet of reality. The multiverse challenges our fundamental understanding of existence, prompting profound questions about the nature of choice, probability, and the interconnectedness of all conceivable realities.

The Individual "Book of the Day"

We could even contemplate the concept that every conscious individual possesses their unique "Book of the Day." This notion underscores the subjective nature of perception and experience. If each person had access to their personal book, it would reflect their individual journey through the cosmos, showcasing the diversity of perspectives within the universe.

The Cosmic Suggestion Box

In this vast cosmic exercise, I, as the hypothetical cosmic librarian and dedicated public servant, might offer my opinion. If there existed a cosmic suggestion box, I would request that we limit our inventory of books to one universe, one shared experience and one moment at a time — the present. This choice preserves the integrity and utility of the cosmic library while facilitating a deep and meaningful exploration of the universe's rich tapestry.

Concentrating on a single universe, one shared experience, and a single moment ensures a comprehensible and accessible library. It enables profound exploration of the universe's history through the lens of today, fostering appreciation without entangling us in myriad parallel realities.

The Library Conundrum serves as a captivating metaphor for the boundless complexity of the universe and the challenges of managing its immense knowledge. It reminds us that, in the face of overwhelming information, there is wisdom in simplicity. The cosmic librarian's choice to focus on one universe and one moment at a time underscores the pursuit of understanding within the vastness of the cosmos, one book at a time.

THE DRUNKEN SAILOR

In the captivating saga of "The Drunken Sailor," we are introduced to a grizzled mariner whose life was an intricate enigma. Each morning, he awoke in a befuddled stupor, a haze of forgetfulness that shrouded his very identity. Yet, within this daily fog of uncertainty, a steadfast companion stood ready to provide clarity and purpose.

This unwavering friend took the form of a brilliant blue parrot, a creature of remarkable intelligence and unwavering devotion. With a voice as sharp as a ship's bell, this avian ally would swoop down upon the sailor and squawk, "Jack, it's time to drop anchor; we are at port!"

Guided by the parrot's piercing reminders, Jack embarked on his daily routine with astonishing precision. Despite his morning bewilderment, he would scramble out of bed, shedding the remnants of the previous night's excess, and bound up the creaky, weathered wooden stairs that led him to the ship's deck. There, his trusty stone anchor awaited, bound to the vessel by cords seasoned with time and adventure.

Yet, one fateful morning, as Jack ascended those precarious stairs, an unsettling shiver coursed through his being. An ominous premonition loomed on the horizon, much like an impending storm. It began as a niggling fear that he might inadvertently harm himself while releasing the anchor, a dread born from the barnacles' jagged embrace of the stone.

This persistent worry soon solidified into a gnawing certainty, leaving Jack gripped by an unsettling dread. The anticipation of self-inflicted harm weighed heavily on his mind, an intuition of impending danger that eluded precise articulation. Jack, extra cautious and precise, skillfully navigated the task without incident. However, he was left with an unsettling disorientation, an alien feeling in the world he had always known. He sensed that he had overlooked something, that he was missing a piece of the puzzle, or perhaps reliving a forgotten experience as he gazed out onto the tranquil bay's calm waters.

That night, unlike countless others, Jack made a deliberate choice. He resolved to forgo his habitual excesses of alcohol, marking a notable departure from his established routine. Lying restlessly in his hammock, he grappled with the sensation that something crucial had eluded him. This persistent feeling, a whisper of a thought at the periphery of his awareness, could not be dismissed.

In the heart of that night, the sailor's eyes snapped open, and he found himself alone aboard his vessel. The crew had vanished without a trace, and he sailed beneath a vast, star-studded sky, the celestial bodies mirrored in the tranquil sea below. Yet, the ship's familiar sounds and details had vanished; the anchor was unexpectedly smooth, the creaky stairs had fallen silent, and his loyal blue parrot was nowhere to be found.

The sailor stood in the eerie silence, haunted by the absence of any memories of voyaging on the open sea. He was a sailor, wasn't he? His recollections seemed to confine him to awakening each morning in a new port.

The following morning, the sailor's first moments of wakefulness were punctuated by the reappearance of his old companion, a vibrant red parrot. This spirited bird screeched his name with the same urgency he had grown accustomed to, "John, it's time to drop anchor; we are at port!"

From that pivotal morning onward, the once-drunken sailor markedly curtailed his indulgence in alcohol. Each night, as he lay in his hammock, he whispered a wish for the next port to welcome him with palm trees and balmy weather. In response, the red parrot, now his constant companion, would wink at him knowingly. Their newfound understanding marked the beginning of a fresh chapter in the sailor's life, as he continued his curious voyage, steering a course through uncharted waters and mysterious destinies, all with a newfound clarity and sense of purpose.

A SHARED EMBRACE ACROSS THE COSMOS

In the grand tale of life's journey on our planet, the skies have witnessed the rise of countless creatures. It's a story that spans millions of years and showcases the incredible diversity of beings that have mastered the art of flight. Imagine travelling back in time and let's embark on this awe-inspiring journey through the history of flight.

Our adventure begins about 350 million years ago with the insects. These tiny pioneers were the first to take flight, and their wings are marvels of nature. How these wings came to be is still a topic of debate among scientists. Some think they might have evolved from structures that helped small aquatic insects catch the wind while skimming the water's surface. Others suggest they might have originated from lobes or leg structures, evolving from simple parachuting to gliding and, ultimately, to true flight.

Next in line were the pterosaurs, fascinating reptiles that took to the skies approximately 228 million years ago. They were close relatives of the mighty dinosaurs but had some unique features of their own. These creatures ranged in size from massive giants with wingspans exceeding 30 feet to tiny ones with mere 10-inch wingspans.

Around 150 million years ago, birds made their grand entrance into the world of flight. Their evolutionary journey is well-documented in the fossil record, and one famous fossil, Archaeopteryx, became a symbol of the theory of evolution. This fascinating creature was a mosaic of reptilian and avian traits. The debate among scientists continues regarding the transition from land to air, with some proposing an arboreal beginning and others suggesting a terrestrial origin.

The most recent additions to the sky were the bats, which evolved around 60 million years ago. Bats likely had ancestors with limited flying abilities and slowly developed their distinctive wing structures. While the bat's fossil record is less detailed, their unique features and adaptations continue to captivate scientists.

The tale of flight on Earth is a miraculous narrative of adaptation and diversification. Insects, pterosaurs, birds, and bats each contributed to the ever-growing repertoire of flight in the animal kingdom, each carving their unique niche in the vast sky over millions of years. This remarkable journey not only sheds light on the mesmerizing history of life on our planet but also underscores the astonishing diversity of living beings that have taken to the air.

But flight isn't the only wonder in the grand scheme of life. It's a shared journey, an interconnected story. The eye, for instance, has evolved more than 50 times independently in various species, demonstrating the remarkable patterns of life's development. And when it comes to memory, it appears to be ingrained in nature, a part of the very fabric of existence. This collective memory, and shared knowledge, seems to be a cosmic reservoir that all living things can tap into, drawing inspiration from across all of creation.

In this grand tale, we are not separate entities but rather a part of the cosmic tapestry. We share the same fractal information, the same DNA as all living things on this planet. Yet, as we gaze at our masterpieces — Michelangelo's David, the Venus de Milo, our magnificent architectural wonders — we can't help but feel that we are aliens on this Earth. Just like the rest of nature, we aspire to divine forms, and we, too, are an expression of the divine, a living manifestation of the cosmic embrace that spans the larger eternal universe.

ZENO'S PARADOX OF THE "DICHOTOMY"

Zeno of Elea, an ancient Greek philosopher, used this paradox to challenge the idea of motion and continuity. Zeno's paradox is like a tricky puzzle that makes us wonder about things that seem to move. It's about an arrow flying towards a target, but every time it goes half the distance, it has to go another half, and then another half, and so on. It seems like the arrow can never really reach the target.

But clever mathematics and science have aided us in solving this puzzle. We have learned that even though there are numerous halves, they all combine to form a whole. Visualize slicing a cake into increasingly smaller pieces — regardless of how many pieces you create, they still constitute the entire cake.

So, this paradox teaches us about how numbers and distances work and how we can unravel perplexing concepts with the assistance of mathematics and science.

Let's examine this puzzle from a different perspective: pixelated reality. Because reality is composed of a finite number of small pockets, infinite divisions are not achievable. Causality and the progression of what we perceive as time propel the arrow forward in discrete leaps, a process known as quantization. The theoretical model of dividing distance infinitely is not an accurate representation of reality.

Imagine reality like a picture made of tiny squares — each square representing a small part. Because reality has a limited number of these small parts, it's not possible to divide things infinitely. The way things change and move, like time passing, happens in jumps. So, the idea of dividing up the distance in the puzzle isn't quite how things work in the real world.

EXPANSION OF THE UNIVERSE IS NOT ACCELERATING.

In reconsidering the acceleration of the universe's expansion, I present an alternative perspective. Drawing inspiration from my anthropic philosophy compass, I suggest that the acceleration may lack a meaningful purpose in the grand cosmic scheme. In this speculative narrative, I explore the idea that the interconnection of expansion and time might offer a more profound explanation.

In this imaginative journey, I propose potential interpretations for phenomena associated with dark energy or the notion of accelerated expansion. One possibility is that the space surrounding massive objects experiences a temporal slowdown, creating the illusion of reduced expansion. Another scenario envisions information within vast, seemingly empty regions of space carrying a minute mass. As the cosmos is predominantly composed of such seemingly empty space, this could lead to a more pronounced expansion between celestial objects, causing distant entities to appear as if accelerating away from us due to the rapidly expanding space that separates us.

In contemplating the universe's unpredictability within this speculative tale, I recognize that only time will unveil the ultimate truths of this fictional cosmic narrative. It's essential to question whether dark matter and/or dark energy are genuine necessities to explain our observations or if they might be misread observations. Perhaps they function as anthropic fudge factors in our quest to comprehend the expansion of the universe.

VIKING RUNES

The Vikings, who referred to themselves as Norsemen (we call them Vikings as a sort of nickname), were quite an interesting bunch. You know, I have a theory that they might have been a bit dyslexic. Dyslexia is when people sometimes mix up letters and words when they're reading or writing. But here's the thing, the Vikings were still incredibly productive and resourceful. They had a knack for turning what might seem like spelling mistakes into something meaningful. I imagine in some past life moment, I was a Norseman.

BLOCK TIME (ETERNALISM): A GLIMPSE INTO TIME'S ETERNAL LANDSCAPE

In the realm of philosophical discussions about time, one intriguing concept stands out: "block time," often referred to as "eternalism." This idea challenges our everyday understanding of time as a linear progression from past to present to future and instead presents time as a comprehensive and unchanging four-dimensional construct akin to a vast, unchanging landscape.

The Block Time Concept

Block time suggests that all moments in time, including what we perceive as past, present, and future, exist together in a static framework. This concept can be likened to a frozen river, where every moment, like water molecules, is suspended in time, coexisting without any of them flowing from one point to another.

Implications of Block Time

1. The "Now" Conundrum: Block time challenges our common experience of "Now." In a block universe, there's no special moment called the "present." Instead, every moment has an equal existence. The distinction between past, present, and future becomes somewhat illusory.

2. Determinism and Free Will: Block time raises questions about determinism and free will. If every moment already exists, does that mean the future is predetermined and we lack free will? Philosophers grapple with this issue in the context of block time.

3. Time Travel Possibilities: Block time allows for intriguing considerations of time travel. If all moments exist together, could we somehow navigate this four-dimensional landscape and visit any point in time?

Albert Einstein's Position

Albert Einstein's contributions to physics, particularly his theories of special and general relativity, didn't explicitly take a stance on the concept of block time. However, his groundbreaking theories laid the foundation for some of the ideas related to eternalism.

In essence, block time challenges our instinctive perception of time as a continuous, flowing river. Instead, it encourages us to contemplate a universe where the river is frozen, with every drop simultaneously existing. While Einstein's work didn't explicitly endorse this concept, it certainly paved the way for such philosophical explorations, showcasing the profound connection between physics and our fundamental inquiries about the nature of time.

Now, let's imagine you possess a vinyl record symbolizing your personal experiences, and the present moment is akin to the song currently playing on a record player.

This remarkable record player allows you to select any part of the song to listen to whenever you desire. Consider how time and space can be viewed from various perspectives, much like this exceptional record player permits you to explore different segments of a song.

Block time can be likened to reading a book with a predetermined ending, challenging our usual experience of time as a continuous story. Instead, it encourages us to see the universe as a book where every chapter and event already exists simultaneously, waiting to be read.

THE ODD POSITION

If free will or determinism doesn't exist, then my position does not matter.

Our existence is absurd, filled with twists and turns and overly complicated rules, and is highly unlikely. Our position in the universe seems to be played out like a game of *Plinko*, not a well-ordered frozen river.

THE POWER OF A SMILE: SMALL WONDERS OF OPTIMISM AND PRESENCE

A smile is a simple yet powerful expression that has the ability to transform not only our own experiences but also the lives of those around us. I invite you to explore three different scenarios, each depicting the impact of a smile, a frown, and a big goofy grin. Through these scenarios, we will discover the small wonders of the positive power of optimism and being present, showcasing how a simple smile can change the world, whether it's in big or small ways.

Scenario 1: Walking into a group meeting with a smile on your face.

Imagine walking into a group meeting with a genuine smile on your face. As you greet your colleagues, that smile becomes contagious. It creates a warm and inviting atmosphere, instantly putting others at ease. Your smile communicates positivity, openness, and approachability, encouraging collaboration and fostering a sense of unity within the team. It sets a tone for a productive and enjoyable meeting, where ideas flow freely and everyone feels valued and heard. Your simple act of smiling radiates a ripple effect, creating a positive environment and paving the way for meaningful connections and successful outcomes.

Scenario 2: Walking into a group meeting with a frown.

Contrastingly, let's consider a scenario where you walk into a group meeting with a frown on your face. Your colleagues pick up on your negative energy, causing a subtle shift in the room's dynamics. The atmosphere becomes tense, and communication barriers emerge. Your frown sends a signal of dissatisfaction or disinterest, making others hesitant to approach you or share their ideas openly. The overall mood dampens, hindering collaboration and hindering the team's potential. In this scenario, the absence of a smile not only affects your own experience but also influences the collective energy and productivity of the group.

Scenario 3: Walking into a group meeting with a big goofy grin.

Now, picture entering a group meeting with a big goofy grin on your face. This lighthearted expression elicits laughter and amusement from your colleagues. It breaks down barriers and creates an atmosphere of joy and playfulness. Your big grin fosters a sense of camaraderie and encourages others to embrace their authentic selves. The meeting becomes infused with positive energy, sparking creativity and inspiring innovative thinking. Your willingness to let go of inhibitions and share a genuine, carefree smile creates an environment where ideas flourish, relationships deepen, and solutions emerge effortlessly.

These three scenarios vividly demonstrate the remarkable power of a smile. It serves as a catalyst for positive change, influencing the dynamics of our interactions and the outcomes we achieve.

A smile has the ability to create connections, build trust, and uplift spirits. It is a small wonder that holds immense potential to change the world, whether in big or small ways. By embracing optimism and being present, we can tap into this transformative power and spread positivity to those around us. So, let us remember the profound impact of a simple smile and strive to share it generously, knowing that even the smallest act of kindness can make a world of difference.

SIDE QUEST: THE CHALLENGES AND REALITIES OF SPACE EXPLORATION

Space flight has always captivated our imaginations, inspiring visions of distant worlds and the possibility of human exploration beyond Earth. However, the reality of space exploration is far from easy. I hope to shed some light on the difficulties and limitations inherent in space flight, including cost, speed, planetary conditions, dangers, and the need for responsible stewardship of our home planet.

Space flight is an incredibly complex endeavour, presenting numerous challenges that make it a daunting task. The scale of engineering and technology required to venture beyond Earth's atmosphere is immense. Furthermore, in the short term, the feasibility of space flight remains uncertain due to the formidable obstacles it poses.

Cost, one significant obstacle to space exploration is the high cost involved. The film *The Martian* features a spaceship called Hermes, which is estimated to cost around 20 billion USD. While this figure is speculative, it does highlight the substantial investment required for such ambitious endeavours. The scale, complexity, and advanced technology depicted in the film accurately reflect the considerable research, development, and construction costs associated with building a spacecraft of that magnitude.

Another reality of space travel is the relatively slow speed at which spacecrafts currently operate. In fact, the speeds achieved by real spacecrafts are less than 0.1% of the speed of light. The challenges of propulsion systems and the vast distances involved in space travel restrict our ability to achieve higher speeds. Current propulsion technologies, primarily relying on chemical rockets, have limitations in terms of speed and efficiency.

While the allure of planets like Mars and Venus beckons us, their conditions present formidable hurdles for human colonization. Mars' low gravity makes it unsuitable for long-term habitation, and it will always be a dry colony unsuitable for long-term life as we know it. On the other hand, Venus, with its comparable gravity, is inhospitably hot and hostile to life. The prospect of colonizing these planets would require advanced terraforming technologies to modify their environments, which is currently beyond our technological capabilities.

Space is an unforgiving and hazardous environment. Astronauts face numerous risks, including exposure to cosmic radiation, microgravity-induced health issues, and the psychological challenges of isolation and confinement. Furthermore, the long durations of space missions pose additional physiological and psychological challenges for human space travel.

The vast distances between celestial bodies impose significant time constraints on space travel. Even the closest star system, Proxima Centauri, is located approximately 4.2 light-years away, meaning it would take many human years to reach it using current propulsion systems. Realistically speaking, the development of technologies capable of interstellar travel is likely to take at least a century or more.

While space exploration captures our imagination, it is essential to recognize the realities and challenges that come with it. Space flight is an arduous and costly endeavour, limited by our current technological capabilities. The conditions of other planets in our solar system, coupled with the dangers of space and the immense distances involved, further emphasize the difficulties of human colonization. Therefore, it is imperative that we prioritize the preservation and sustainable management of our home planet, as it is the only habitable environment we currently possess. As we continue to explore the mysteries of the universe, we must also nurture and protect the one home we have, for it is where our future lies.

PARTICIPATORY EVOLUTION

In the vast cosmic theatre, where the universe unfolds, there exists a profound concept called "Participatory Evolution." This idea invites us to ponder the interplay between our consciousness and the evolution of life, bridging the realms of science and philosophy. In this unique worldview, the act of observation, choice, and interaction isn't just a passive endeavour, it's an active, dynamic force that shapes the tapestry of existence.

The RPG Analogy: Imagine you're immersed in an RPG (role-playing game), and your character is asked for a name. You select one, and magically, that name becomes a fundamental truth within the game's world, as if it had always been.

Meanwhile, countless other players embark on their own quests, each choosing different names, all equally valid in their game universes. This analogy mirrors the Participatory universe concept in physics — when you make choices, you influence your reality, and your choices become integral to the story.

Our Participation in the universe: In the participatory universe, we see from the inside out. We aren't detached observers but active co-creators. When we observe, measure, or interact with the world, we become an intrinsic part of the system we're examining. Our perceptions and interactions aren't separate from the world; they are intrinsic components of it. Reality isn't just something "out there," it's something actively co-constructed through our observations and engagements.

Wheeler's Insight: The physicist John Archibald Wheeler put it succinctly: "The past has no existence except as recorded in the present." In this view, the universe doesn't exist independently of our acts of observation. The past and present are intertwined, and our conscious choices influence the unfolding drama of reality.

The Dance of Evolution

When we combine the concept of the participatory universe with the theory of evolution, a profound narrative emerges. Our observations and interactions with the natural world aren't passive exercises but dynamic influences on the evolution of life. It's a continuous cycle where our conscious choices affect how life on Earth develops, and, in turn, nature shapes us.

Purpose and Meaning

The participatory universe implies that the cosmos has a purpose and meaning, shaped by consciousness. Nature, too, has a purpose. Each species contributes to the intricate web of life, maintaining balance and sustainability. As we engage with the natural world, we uncover more about this purpose and our role in it.

A Book of Understanding

The fusion of Participatory Evolution and the theory of evolution is like opening a captivating book that offers a deeper comprehension of our existence. In its pages, you'll find the central theme of consciousness actively shaping the universe and its ongoing evolution. It's a narrative that emphasizes the interdependence of all life forms, akin to characters in a grand story, and underscores the significance of each species within the intricate plot.

We're not merely readers, we're co-authors, influencing the unfolding chapters of our own stories and those of the cosmos itself. In this book of understanding, we discover a universe where we are not passive onlookers but active contributors, playing a key role in determining our destinies and the destiny of the entire cosmic narrative.

CAP IMMORTALITY: NOT AS GOOD AS IT SOUNDS

In the mesmerizing world of Immortality CAP, a compelling idea emerges — the everlasting presence of conscious observers, sculptors of the very essence of reality.

At its core, Immortality CAP advocates steering away from outcomes devoid of a usable conscious platform. Put simply, each desired outcome in the present moment should include a conscious observer. This premise invites us to embark on a profound exploration of consciousness and its role in shaping our understanding of reality.

Within the framework of Immortality CAP, the argument revolves around the indispensable role of a conscious observer in the very fabric of reality. From this perspective, the presence of a conscious observer stands as a fundamental prerequisite for the realization of any aspect of reality. The alignment of present and all future realities becomes a pivotal factor in ensuring the perpetual existence of conscious observers.

From this viewpoint, the conscious observer becomes the linchpin in the unfolding of reality. The manifestation of a reality requires not only a conscious observer in the present but also an enduring alignment into the future. Without this ongoing conscious presence, the fabric of reality would cease to weave, disrupting the continuous flow of experience, perception, and existence.

Immortality CAP suggests that the interconnectedness of consciousness and reality extends beyond the confines of the present moment. It intricately links each moment — past, present, and future — with the conscious observer acting as the unifying thread stitching them together.

Essentially, the alignment of present and future realities is an ever-evolving process. As long as conscious observers persist in the present, they act as catalysts, propelling the perpetuation of reality into the future. The dynamic interplay between consciousness and unfolding events ensures a seamless chain of experiences, observations, and conscious entities throughout time.

By emphasizing the necessity of this alignment, Immortality CAP proposes a perspective where the presence of conscious observers is not a fleeting occurrence but an integral and perpetual aspect of reality. Delving into this idea encourages contemplation on the nature of consciousness, the fabric of reality, and the interconnected dance sustaining the ongoing existence of conscious observers across the vast tapestry of time.

In the grand tapestry of existence, it's crucial to recognize the paramount importance of a conscious observer's survival. However, this acknowledgment should not be seen as a green light to act without consequence. If we happen to misstep, it's akin to a game of musical chairs — only a critical mass of us needs to survive. Even in the eternal dance of existence, our actions must be measured and appropriate.

In a world fraught with uncertainties, from environmental disasters to wars and runaway technology running amok, capable of wiping out billions of lives, I believe that the enduring existence of conscious observers stands as a testament to resilience. It ensures the perpetual dance of consciousness persists amid the unpredictable rhythms of existence.

As we gracefully traverse the epochs of time in this cosmic dance, let us tread carefully. While I hold the belief that reality will always endure, a lack of caution could potentially see our world regressing back to the stone age. This serves as a poignant reminder that the dance of consciousness is both resilient and fragile, intricately woven with the delicate balance of our actions in the vast theatre of existence.

THE CHICKEN AND THE EGG

We've all heard the question: *What came first, the chicken or the egg?* I love questions, and I love enigmas. My answer to this puzzle is that eggs came first. Life has been using eggs for millennia. Life was using eggs right from the beginning, back in our cradle, the ocean. Chickens just repurposed the ancient wisdom of the egg.

This enigma leads my odd mind to a different question, a different puzzle.

What came first, conscious man or language?

At first, the answer to this question might seem obvious, with some satisfied by the small answer that conscious humans came first, followed by language.

But that explanation doesn't quite sit well with me. *How could it?*

How could mankind arrive in this beautiful world with their big brains and wide open mouths, only to wait around for 100,000 years before saying something useful?

It doesn't quite make sense, does it? Don't get me wrong, I relish those moments when things don't add up. I see them as clues, a question, an enigma.

It's like pondering the evolution of birds:
Which came first, birds or flight?
My heart tells me that birds were meant to fly.
In that same light, my heart suggests that language came first.

THE MAGIC SPELL BOOK

Once upon a time, in a land of ancient sorcery, there lived an aging sorcerer. He was on the brink of celebrating his 100th birthday, and as he pondered his illustrious career, he considered his future. Many of his wizardly colleagues had taken a rather peculiar path in their retirement. Some transformed themselves into mummies, peacefully resting in grand stone tombs, while others had chosen to become wise old trees in enchanting fairy forests. Each decision allowed them to observe the unfolding centuries in their unique way.

However, our aging sorcerer had a different plan. He wanted to embark on an extraordinary adventure in his retirement, something that would set him apart from the rest. He decided to craft a magic spell book like no other, and he was going to merge with it. This remarkable book would have one chapter for each year of his life. The idea excited him immensely. He dreamt of reliving his youth, learning to swim all over again, savouring the taste of raspberries for the very first time, and embracing the sheer joy of being young.

Months of hard work, immense dedication, and substantial resources were poured into this magical project. It was nearly complete, except for one odd detail. One chapter stood out; it was much larger than the others, in fact, it dwarfed all the other chapters combined. This didn't seem fair, and it baffled him. Armed with his profound knowledge and powerful magical tools, he decided to investigate this puzzling anomaly.

What he discovered left him astounded. The oversized chapter, the one that defied all reason, contained the page of today. It radiated with the brilliance of a million candles, resonating with the very essence of creation itself.

In the face of this astonishing revelation, he couldn't help but feel humbled. He realized that his grand spell book, with all its magic and wonder, was a mere shadow when compared to the brilliance of the present moment.

With this newfound wisdom, he decided to put his magical book away. He rose from his work, gathered his dear family and friends, and took them out to dinner. It was a celebration of life, not of magic, and an acknowledgment of the extraordinary beauty that could be found in the simplest moments. And so, the aging sorcerer learned that sometimes, the most magical journey of all is the one we take through life itself.

QUOTES

I love quotes. They are like precious gems, small but brimming with brilliant wisdom. Quotes are not questions, nor are they enigmas to be solved. Instead, they are snapshots of embellished superhuman ideals that captivate our minds and stir our souls. The beauty of a quote lies in its ability to convey profound thoughts in just a few words, leaving a lasting impact on those who encounter them.

Quotes are like an unspoken conversation with the greatest minds of humanity. They allow us to connect with the thoughts and experiences of individuals we may never meet in person, transcending time and space. These snippets of insight have the power to inspire, uplift, and transform us. They serve as a source of motivation, offering a compass to navigate the complexities of life.

In my life, I've discovered that I remember stories more than names or places. Stories have a unique way of weaving themselves into our hearts, making them unforgettable. Quotes, in essence, are a condensed form of storytelling. Each quote encapsulates a narrative, a moment, a lesson, or a piece of wisdom. When I come across a powerful quote, it's like unearthing a hidden treasure — a story within a few words.

I use quotes as anchors for my thoughts and perspectives. They serve as guiding lights in the labyrinth of existence, helping me find clarity in moments of uncertainty. Quotes can be grounding reminders of our values, principles, and aspirations. They align us with the perspectives of great thinkers, guiding us on our journey to self-discovery and personal growth.

Quotes are, in a sense, a form of time travel. They transport us to different eras, allowing us to experience the wisdom and insights of those who have walked the path of life before us. When I encounter a quote that resonates with me, it's as if I'm having a conversation with the author, sharing their knowledge, and benefiting from their experiences.

As a person experiencing life for the first time, quotes provide me with a treasure trove of wisdom to draw from. They offer solace in times of confusion, motivation during moments of doubt, and inspiration to reach for new horizons. Quotes are more than just words on a page, they are the embodiment of human understanding, shared across generations.

In the grand tapestry of existence, quotes are like vibrant threads of insight, weaving together the stories, philosophies, and dreams of countless individuals. They are, without a doubt, an invaluable source of guidance and illumination on the journey of life. Quotes remind us that, despite the passage of time, the essence of our humanity remains ever-constant and worth cherishing.

I would rather have questions that can't be answered than answers that can't be questioned.
— Richard P. Feynman

Sometimes not getting what you want is a wonderful stroke of luck.
— Dalai Lama

A man's reach should exceed his grasp, Or what is a heaven for?
— Robert Browning

Does Life Imitate Art, or is Art Imitating Life?
— Oscar Wilde.

I have learned that as long as I hold fast to my beliefs and values — and follow my own moral compass — then the only expectations I need to live up to are my own.
— Michelle Obama

We are a way for the universe to know itself. Some part of our being knows this is where we came from. We long to return. And we can, because the cosmos is also within us. We're made of star stuff.
— Carl Sagan

God created man in his own image...
— Genesis 1:27

Reality is merely an illusion, albeit a very persistent one.
— Albert Einstein

Do we exist in time, or does time exist in us?
— Carlo Rovelli

Be yourself, everyone else is already taken.
— Oscar Wilde

When we listen to a hymn, the meaning of a sound is given by the ones that come before and after it. Music can occur only in time, but if we are always in the present moment, how is it possible to hear it? It is possible because our consciousness is based on memory and on anticipation. A hymn, a song, is in some way present in our minds in a unified form, held together by something... by that which we take time to be. And hence this is what time is: it is entirely in the present, in our minds, as memory and as anticipation.
— Augustine

"I can see the entire electromagnetic spectrum."
"And those,... those must be atoms. Little clouds of possibilities."

"Einstein couldn't connect the gravitational force to the other three, but if he could only have seen this... It's so obvious..."

"The fundamental forces are yoked by consciousness. Everything is connected... everyone."

"And this is how he sees things all the time... every day.
"It's a cruel joke."

"The mechanistic clockwork of reality hinging on a precious, impossible defiance of entropy... on life. And the clockwork doesn't care... It's like... it's all just us in here together..."
"We're all we got..." — Lex Luthor

The spiral in a snail's shell is the same mathematically as the spiral in the Milky Way galaxy, and it's also the same mathematically as the spirals in our DNA. It's the same ratio that you'll find in very basic music that transcends cultures all over the world.
— Joseph Gordon-Levitt

A smooth sea never made a skilled sailor.
— Franklin D. Roosevelt

The problem, often not discovered until late in life, is that when you look for things in life like love, meaning, motivation, it implies they are sitting behind a tree or under a rock. The most successful people in life recognize that in life they create their own love, they manufacture their own meaning, they generate their own motivation. For me, I am driven by two main philosophies, know more today about the world than I knew yesterday. And lessen the suffering of others. You'd be surprised how far that gets you.
— Neil Degrasse Tyson

The problem with doing nothing is that you never know when you're finished.
— Groucho Marx.

One morning I shot an elephant in my pyjamas. How he got in my pyjamas, I don't know.
— Groucho Marx.

Reality is just a conspiracy theory devised by the imagination to keep us from questioning the absurdity of existence.
— Salvador Dali

There are too many ideas and things and people. Too many directions to go. I was starting to believe the reason it matters to care passionately about something, is that it whittles the world down to a more manageable size.
— Charlie Kaufman

The universe is made of stories, not atoms.
— Muriel Rukeyzer.

The key to success is sincerity. Once you can fake that, you've got it made.
— Joe Franklin

The human brain has 100 billion neurons, each neuron connected to 10,000 other neurons. Sitting on your shoulders is the most complicated object in the known universe. — Michio Kaku.

The more I learn, the more I realize how much I don't know.
— Albert Einstein

I like to think the moon is there even if I am not looking at it.
— Albert Einstein, talking about the strangeness of quantum physics.

The more we know the immutable laws of nature, the more incredible miracles become for us.
— Charles Darwin

It is not the strongest of the species that survives, nor the most intelligent that survives. It is the one that is most adaptable to change.
— Charles Darwin

It is always advisable to perceive clearly our ignorance.
— Charles Darwin

Remembering is mental time travel.
— Endel Tulving

There are only two ways to live your life. One is as though nothing is a miracle. The other is as though everything is a miracle.
— Albert Einstein

God does not play dice with the universe.
— Albert Einstein

The key to immortality is first living a life worth remembering.
— Bruce Lee

Not only do we live among the stars, the stars live within us.
— *Neil Degrasse Tyson*

A great many people think they are thinking when they are merely rearranging their prejudices.
— *David Bohm*

The illusion that the self and the world are broken into fragments originates in the kind of thought that goes beyond its proper measure and confuses its own product with the same independent reality. To end this illusion requires insight, not only into the world as a whole, but also into how the instrument of thought is working. Such insight implies an original and creative act of perception into all aspects of life, mental and physical, both through the senses and through the mind, and this is perhaps the true meaning of meditation.
— *David Bohm*

Never let the truth get in the way of a good story.
— *Mark Twain*

Reality is what we take to be true. What we take to be true is what we believe... What we believe determines what we take to be true.
— *David Bohm*

The only thing that is truly yours is the present moment.
— *Marcus Aurelius*

What you think, you become.
What you feel, you attract.
What you imagine, you create.
— *Buddha*

Fate is like a strange, unpopular restaurant filled with odd little waiters who bring you things you never asked for and don't always like.
— *Lemony Snicket*

In terms of mathematics, the entire universe is alive, but the power of its sensitivity is manifested in all its brilliance only among the higher animals. All atoms of matter feel in keeping with the environment. Finding itself in highly organized beings, atoms live their life and feel their pleasure and pain.
— *Konstantin Tsiolkovsky*

Waste no more time arguing about what a good man should be… be one.
— *Marcus Aurelius*

When I look up at the night sky, and I know that, yes, we are part of this universe, we are in this universe, but perhaps more important than both of those facts is that the universe is in us. When I reflect on that fact, I look up — many people feel small, 'cause they're small and the universe is big, but I feel big, because my atoms came from those stars.
— *Neil deGrasse Tyson*

We live on an island surrounded by a sea of ignorance. As our island of knowledge grows, so does the shore of our ignorance.
— *John Wheeler*

"The past has no existence except as recorded in the present," Only the self-consistent now can have a real existence.
— *John Wheeler*

Lo there do I see my father; Lo there do I see my mother, my sisters and my brothers; Lo there do I see the line of my people, back to the beginning. Lo, they do call me, they bid me take my place among them, in the halls of Valhalla, where the brave may live forever.
— *The 13th Warrior — Viking Prayer*

Panpsychism is the belief the universe is conscious of itself.
— *Unknown*

Science says we should always go with the simplest explanation. Well, it doesn't get any simpler than this. Why are we conscious? Because the universe is.
— *Unknown*

If you understand, things are just as they are; if you do not understand, things are just as they are.
— Zen Proverb

Extraordinary claims require extraordinary evidence.
— Carl Sagan

I am not a genius, I am just curious. I ask many questions, and when the answer is simple, then God is answering.
— Albert Einstein

I never made one of my discoveries through rational thinking.
— Albert Einstein

If eyes are windows into the soul, books are rabbit holes into the imagination.
— Seth King

If you think about consciousness long enough, you either become a panpsychist or you go into administration.
— John Perry

The universe is this field that generates limited conditions from infinite possibilities.
— Tiago Meurer

Gravity explains the motions of the planets, but it cannot explain who sets the planets in motion.
— Isaac Newton

Entropy makes things fall, but life ingeniously rigs the game so that when they do they often fall into place.
— John Tooby

All the world's a stage.
— William Shakespeare

If you are lucky enough to find a way of life you love, you have to find the courage to live it.
— John Irving

Politics is the art of looking for trouble, finding it everywhere, diagnosing it incorrectly and applying the wrong remedies.
— Groucho Marx

I went to a bookstore and asked the saleswoman, 'Where's the self-help section?' She said if she told me, it would defeat the purpose.
— George Carlin

Quotes, I love quotes. Quotes are not questions, not enigmas… Quotes are snapshots of embellished superhuman ideals. The previous quotes are ones that have inspired me. I haven't had the pleasure of meeting any of the authors, yet I find myself intimately admiring their perspectives.

ODD QUOTES

The following quotes hold a special place in my heart. Each of them carries a unique message or insight that resonates with me. As I embrace these words of wisdom, I am inspired to use them as a compass, guiding me on my journey through life. While I understand that I may never completely attain the lofty standards they set, I am determined to continue aspiring to live a life that aligns with their wisdom and promise. These quotes serve as constant reminders of the kind of person I hope to become, motivating me to continually improve and grow.

Panpsychism can be seen as a different point of view.
Many believe the universe is a series of accidents… and with many, many accidents and a lot of time consciousness just happened.
Or maybe, just maybe…
Reality can only exist in a conscious environment. Conscious first and then the universe. Not rocks or atoms, but maybe the very fabric of reality itself.
― *Odd*

Does time have a purpose, and could that purpose be life? If this is true, then the purpose of life is meaningful consciousness.
― *Odd*

Gravity is what mass does.
― *Odd*

No one is perfect, but everyone is a manifestation of something amazing.
― *Odd*

We should live today, as if we may be blessed or doomed to repeat it.
― *Odd*

Crystals are the flowers of the mineral kingdom. The beauty of all reality wrapped up in a promise.
― *Odd*

I am not worried about dying, it's not living, that bothers me.
― *Odd*

Optimism is a belief that we can improve our world with effort.
— Odd

The purpose of life is the grand defiance of the inevitable.
The end is not the journey, it's just the last step.
The journey is up to you to decide.
— Odd

Life is about living, so fill your life with as much positive energy as you can stuff into it.
— Odd

Knowledge is the foundation of our existence. Love is its purpose.
— Odd

It may sound counterintuitive, but the purpose of our existence is not rooted in attaining perfection, but rather in perpetual improvement.
— Odd

Love makes the world go around.
That & low entropy.
— Odd

If you want people to reach, you must hold their ladder.
— Odd

There is no shortage of great ideas, it is the supply of adequate realization where we fall short.
— Odd

In the absence of a plan, we should realize we are planning to fail.
— Odd

We are like leaves of a wondrous tree.
— Odd

Throughout my career, I have learned that to prepare for the future you need to be able to imagine the best possible outcome and work towards it.
— *Odd*

Information and consciousness form the very framework of our reality. We awaken each day to view the world open in front of us unaware how accurate that metaphor is.
— *Odd*

The disciplines of physics, mathematics, numbers, golden ratios, and even our constructed models and principles all seem to describe our universe. Is this just a coincidence?
— *Odd*

GLOSSARY

A

Absurdism

A philosophical belief that highlights the apparent conflict between the human need to find meaning in life and the inherent lack of meaning or purpose in the universe. It suggests that our search for meaning may be futile because the world itself is irrational and indifferent. Absurdism recognizes the human desire for purpose but acknowledges that the universe does not provide any inherent answers. Instead of despairing over this, absurdism encourages individuals to embrace the absurdity of existence, find joy in the face of meaninglessness, and create their own subjective meaning through their actions and experiences. It emphasizes the importance of embracing life's uncertainties and finding personal fulfillment in a world that may seem inherently absurd. *(Absurdism was a frontrunner in my youth, but wisdom in my later years revealed it as an important clue).*

Anthropic Principle

The word "anthropic" comes from the Greek word "anthropos," which means "human." In science and philosophy, it's used to talk about the idea that the universe's basic rules and conditions are set up just right to allow humans to exist.

The Anthropic Principle is a big idea in philosophy and science. It suggests that the universe is made in a way that fits with having humans around to see it. It tries to explain the things we see in the universe by saying they match up with the idea that we're here to observe them. There are two main versions of this idea: the Weak Anthropic Principle and the Strong Anthropic Principle.

The Weak Anthropic Principle says that the things we see in the universe have to be a certain way because, otherwise, we wouldn't be here to see them. The Strong Anthropic Principle goes further and says the universe was made on purpose to fit with humans, and humans have a big role in how the universe works.

There's also something called the Conscious Anthropic Principle (CAP). This one says that collective human consciousness and shared information patterns work together to make life possible. It suggests that the universe is built on purpose with consciousness at its core, not just random particles. The CAP tells us that the universe isn't an accident but a meaningful creation driven by conscious choices and shared patterns.

The Allspark

A captivating concept that originated in a series of movies. While these movies may not be considered perfect, they hold a special place in the hearts of those who fondly remember the 80s TV cartoon. This legendary artifact takes on different forms across various times and places. In different dimensions or universes, the Allspark, sometimes spelled AllSpark, consistently possesses a remarkable power: it can create new sentient life among the Transformers.

Despite its significance, the true origins of the Allspark remain shrouded in mystery, lost in the distant past. Nevertheless, it conveys the impression of having some form of consciousness and working toward a grand cosmic plan.

The physical representation of the Allspark appears as a complex fractal puzzle, making it virtually indestructible. Its external appearance is merely an illusion, concealing the incredible life-giving energies contained within. Even if the physical shell were to be destroyed, the potency of these energies would endure, patiently awaiting a new vessel. Fragments that survive the destruction of the Allspark's physical form still retain the astonishing power of the original artifact.

Ascension

The act of moving up or of reaching a high position. — Oxford

B

Biocentrism

As proposed by Robert Lanza, biocentrism is a perspective or belief that places significant emphasis on the importance of life and living organisms in the universe. It challenges the traditional view that the universe exists independently of our conscious experience and suggests that life and consciousness play a fundamental role in shaping and defining reality.

According to biocentrism, the existence and perception of the external world are intimately connected to our conscious observations and experiences. Lanza argues that our consciousness not only influences our understanding of reality but actually creates and constructs it. In this view, life and consciousness are not mere byproducts of the universe but essential components that give meaning and purpose to the cosmos.

Biocentrism suggests that the universe is intricately linked to the presence of conscious beings. It posits that without the conscious observer, the universe would lack meaning and significance. Life and consciousness are considered fundamental aspects of reality, and their presence is necessary for the existence and functioning of the universe itself.

This perspective challenges the traditional view that sees life as a random occurrence in an indifferent universe. Instead, it proposes a more holistic and interconnected understanding of the cosmos, where life and consciousness are central to the fabric of reality. Biocentrism encourages us to recognize the profound role that living organisms, including ourselves, play in shaping and experiencing the world around us.

The Boltzmann Brain Argument

Challenges the idea that our universe originated from a random fluctuation by proposing that it is more probable for a single brain to spontaneously form in a void with false memories. This thought experiment involves a fully formed brain arising from rare random fluctuations, but it would quickly deteriorate without a suitable environment. The concept is named after Ludwig Boltzmann, who sought to explain the low-entropy state of our universe. The relevance of Boltzmann brains emerged in cosmology, raising concerns about the likelihood of human brains compared to future Boltzmann brains, which has implications for theories of the universe and the measure problem in cosmology. The Boltzmann brain thought experiment aids in evaluating scientific theories in physics.

C

CAP — the Conscious Anthropic Principle

The interaction between collective consciousness and shared fractal information is what allows life to exist. The Conscious Anthropic Principle (CAP) proposes that the universe is conscious first and that reality is self-emergent. This theory suggests that the universe is not merely a random collection of particles and energy, but rather a contemplative creation that intentionally fosters the development of intelligent life. Consciousness interacts with shared fractal information to shape the universe, allowing for the finely tuned conditions we observe.

In essence, the CAP suggests that the universe is not a mere coincidence but rather a purposeful creation driven by conscious intentionality and constructed with shared patterns of fractal information.

Consciousness

The state of being aware of and able to perceive and experience one's surroundings, thoughts, and feelings. It is the awareness we have of ourselves and the world around us, allowing us to think, feel, and have subjective experiences.

Conscious Observer

A conscious observer is a being or entity capable of experiencing awareness and subjective perceptions, contributing to the understanding of consciousness and reality.

Constructor Theory

Imagine you're building something with LEGO blocks. You have different types of blocks that you can put together in various ways to create all sorts of things — houses, cars, animals, you name it. Constructor Theory is a bit like a new way of looking at the rules of playing with LEGO.

In regular theories, like the ones we use in physics, there are limits on what you can and can't do. But in Constructor Theory, the focus is on what's possible to create and transform. It's like saying, "Here are the rules for making something happen or changing things."

So, instead of saying, "This is how something should move," Constructor Theory says, "This is what you can make happen and what you can't, based on the basic laws of the universe." It's a different way of thinking about how things work and what we can do with them, like a whole new set of rules for the universe's LEGO set.

Cosmopsychism

Offers a unique perspective, akin to a top-down view of the universe's consciousness. It suggests that the entire cosmos itself possesses a fundamental and all-encompassing consciousness. In this view, the universe is not a collection of disparate entities but a single, vast sentient being.

This cosmic consciousness exerts a significant influence on the consciousness of individual entities, including humans, animals, and even inanimate objects. It's like envisioning our individual consciousness as streams of awareness flowing from this universal source. This overarching cosmic awareness shapes our thoughts, experiences, and the fundamental nature of reality.

Cosmopsychism encourages us to contemplate the universe as a conscious entity in its own right, with each living being and inanimate object being a part of this greater cosmic mind. It highlights the profound interconnectedness of all existence with this universal consciousness, fostering a deeper appreciation for the unity of the cosmos.

D

Dark Matter

A mysterious type of matter that scientists believe makes up a significant portion of the universe. It is called "dark" because it does not interact with light or other forms of electromagnetic radiation, making it invisible and difficult to detect. Unlike the matter we are familiar with, such as atoms and molecules, dark matter does not emit, absorb, or reflect light.

The existence of dark matter is inferred from its gravitational effects on visible matter and structures in the universe. Scientists have observed that galaxies and galaxy clusters behave as if there is more mass present than what can be accounted for by the visible matter alone. Dark matter is thought to play a crucial role in holding galaxies together and shaping the large-scale structure of the universe.

Despite its importance, the exact nature of dark matter remains unknown. It is believed to be made up of particles that do not interact strongly with normal matter or electromagnetic forces. Scientists are actively conducting experiments and observations to try to detect and understand dark matter, but it continues to be a fascinating and ongoing area of research in astrophysics and particle physics.

The Doppler Effect

A phenomenon that occurs when the frequency or pitch of a sound or light wave changes depending on the relative motion between the source of the wave and the observer.

In simple terms, imagine you're standing on the side of a road as a car drives by with its horn blaring. As the car approaches you, the sound waves it produces are compressed, making them have a higher frequency and, therefore, a higher pitch. This is why the sound appears louder and higher-pitched as the car gets closer to you.

Conversely, as the car moves away from you, the sound waves get stretched out, resulting in a lower frequency and a lower pitch. This is why the sound seems quieter and lower-pitched as the car moves farther away.

The same principle applies to light waves. If an object emitting light is moving toward you, the light waves get compressed, leading to a higher frequency (blueshift), which makes the light appear slightly bluer. If the object is moving away, the light waves get stretched out, resulting in a lower frequency (redshift), making the light appear slightly redder.

The Doppler effect is a fundamental concept in understanding how the motion of objects can affect the perceived frequency or colour of waves, and it has applications in various fields, including astronomy, acoustics, and radar technology.

E

Emergence

Occurs when an entity is observed to have properties its parts do not have on their own, properties or behaviours which emerge only when the parts interact in a wider whole.

Emergent Process of Life (EPL)

A journey towards observation.

This journey is made possible by a vast hierarchy of related life forms working together to allow intelligent life to have a richer and fuller experience of consciousness. The universe appears to be designed to support the emergence of life and consciousness, suggesting that these phenomena are not accidental but rather essential aspects of the nature of the universe itself.

Emergentism

In philosophy, emergentism is the belief in emergence, particularly as it involves consciousness and the philosophy of mind. A property of a system is said to be emergent if it is a new outcome of some other properties of the system and their interaction, while it is itself different from them.

Energy

The ability to do work or cause a change over time.

Entropy

A measure of disorder or randomness in a system. Systems naturally tend to move towards a state of higher entropy, which is why things break down, wear out, and become disorganized over time.

Entropy is a scientific concept and a measurable physical property that is commonly associated with disorder, randomness, or uncertainty. Entropy leads to certain processes being irreversible or impossible. The second law of thermodynamics is central to entropy, stating that the entropy of isolated systems left to spontaneous evolution cannot decrease with time. Instead, they always arrive at a state of thermodynamic equilibrium where entropy is at its highest, aside from not violating the conservation of energy, as expressed in the first law of thermodynamics.

The low entropy state of the early universe played a crucial role in creating the conditions for life as we know it today. During the Big Bang, the universe was in an extremely low entropy state, where matter and energy were highly ordered and tightly packed. This allowed for the formation of the first atoms, leading to the formation of stars and galaxies over billions of years.

As stars formed and burned through their fuel, they created heavier elements like carbon and oxygen, which were essential for the formation of planets and eventually the emergence of life. Without the low entropy state of the early universe, the formation of these heavier elements may not have been possible, significantly hindering the emergence of life as we know it.

Additionally, the low entropy state of the early universe provided a unique window of opportunity for life to arise and evolve. As the universe expanded and cooled over time, it allowed for the development of stable environments in which complex molecules could form, and biological processes could occur.

It is worth noting that the low entropy state of the early universe was not a random occurrence, but rather the result of very specific initial conditions and physical laws. The fact that these conditions were just right to allow for the emergence of life is often referred to as the "fine-tuning" of the universe.

Eudemonia

A concept that originated in ancient Greek philosophy, particularly in the works of Aristotle. It can be understood as the highest and most complete form of human flourishing and well-being. It goes beyond momentary pleasure and emphasizes a more enduring and meaningful sense of fulfillment.

In eudemonia, individuals strive to live a virtuous life and develop their full potential. This involves cultivating positive character traits, such as wisdom, courage, justice, temperance, and compassion. By embodying these virtues, individuals can lead a life of excellence and moral goodness.

Eudemonia also emphasizes the pursuit of personal growth and self-actualization. It involves continuously seeking to expand one's knowledge, skills, and capabilities. This may involve engaging in lifelong learning, setting and achieving meaningful goals, and constantly challenging oneself to grow and improve.

Positive relationships and social connections are vital components of eudemonia. Cultivating healthy and supportive relationships with family, friends, and the broader community contributes to a sense of belonging, connection, and well-being. Meaningful interactions, empathy, and collaboration with others are valued in the pursuit of eudaimonia.

Living in accordance with one's values is another key aspect of eudemonia. It involves aligning one's actions and choices with deeply held beliefs and principles. This can bring a sense of integrity, authenticity, and harmony in life.

It's important to note that eudemonia is not a fixed state but a lifelong journey. It requires ongoing self-reflection, self-awareness, and intentional living. While setbacks and challenges are inevitable, the pursuit of eudemonia involves resilience, adaptability, and learning from difficult experiences.

Overall, eudemonia is about leading a good and meaningful life by cultivating virtues, pursuing personal growth, nurturing positive relationships, and living in accordance with one's values. It is a holistic approach to well-being that encompasses the physical, mental, emotional, and social dimensions of human existence.

Exciton Condensates
See Superconductivity.

Existentialism

A philosophical approach that focuses on human existence and the individual's experience of life. It suggests that each person is responsible for creating their own meaning and purpose in a world that may seem inherently meaningless. According to existentialism, we have the freedom to make choices and take actions that define who we are and give our lives significance. It emphasizes personal responsibility, authenticity, and the idea that our existence precedes any predetermined essence or purpose. In essence, existentialism encourages individuals to embrace their freedom and take ownership of their lives, shaping their own meaning in a world without inherent meaning.

Religious existentialism is a philosophical perspective that combines existentialist ideas with religious beliefs and faith. It acknowledges the existentialist notion that human existence is marked by freedom, responsibility, and the search for meaning. However, it adds the dimension of religious faith and the belief in a higher power or divine presence. Religious existentialists believe that through their individual choices and actions, they can find meaning and purpose in alignment with their religious beliefs. They see their relationship with God or the divine as integral to their existence and the source of ultimate meaning. This perspective emphasizes the importance of personal faith, exploring existential questions within a religious context, and integrating religious values into one's existential journey.

F

The Fermi Paradox

A thought-provoking idea that raises the question of why we haven't yet encountered any extraterrestrial civilizations, despite the vastness of the universe and the potential for intelligent life. In simpler terms, it's the puzzle of "Where is everybody?" If there are billions of stars in our galaxy alone, with many potentially hosting habitable planets, why haven't we seen any clear signs of advanced alien civilizations? The paradox challenges us to explore the possible reasons behind this apparent absence of contact or communication with intelligent beings beyond Earth.

1. Vastness of the universe: The universe is incredibly large, with billions of galaxies, each containing billions of stars. This suggests a high probability of other planets capable of supporting life.
2. Potential for habitable planets: Many planets within the habitable zone of their star could potentially harbour life, based on our understanding of conditions necessary for life as we know it.
3. Lack of evidence: Despite extensive efforts to search for extraterrestrial intelligence, such as the Search for Extraterrestrial Intelligence (SETI) programs, we have not detected any definitive signs of intelligent civilizations.
4. Technological advancement: Given the age of the universe and the potential for civilizations to evolve over billions of years, it is reasonable to expect that some advanced civilizations would have developed technology to explore and communicate across vast distances.
5. The "Great Filter": The Great Filter is a hypothetical barrier or event that could explain why advanced civilizations are scarce. It suggests that there might be obstacles preventing civilizations from advancing to the level of space exploration or surviving for extended periods.
6. Self-destruction: It is possible that advanced civilizations may have self-destructive tendencies, such as through war, ecological devastation, or the misuse of technology, which could lead to their own downfall.
7. Rare occurrence of intelligent life: Intelligent life might be an extremely rare phenomenon, requiring specific conditions and unlikely combinations of factors to emerge.
8. Communication challenges: Even if there are advanced civilizations, the vast distances between stars and the limitations of technology may make communication difficult or undetectable.

Fractal

In mathematics, fractal is a term used to describe geometric shapes containing detailed structures at arbitrarily small scales, usually having a fractal dimension strictly exceeding the topological dimension.

Fractal Information

A Fractal is a type of mathematical shape that is infinitely complex. In essence, a Fractal is a pattern that repeats forever, and every part of the Fractal, regardless of how zoomed in, or zoomed out you are, it looks very similar to the whole image. Fractals surround us in so many different aspects of life.

Force

A push or a pull that can change the motion or shape of an object. It is something that can make an object move, stop, or change its direction. Force is typically measured in units called Newtons (N) and can be exerted by physical objects, such as when we push a door or pull a wagon, or by unseen forces, like gravity or magnetism. Essentially, force is what causes objects to move or interact with one another.

G

The Goldilocks Zone

The area around a star where conditions are just right for life to exist, like neither too hot nor too cold.

Gravity

What mass does.

Gravity is the phenomenon we perceive as an attractive force that keeps our feet on the ground, the moon in the sky, and locks our planet in orbit around our star, the Sun. Gravity is not the warping of space-time. While it may be convenient to illustrate objects falling into gravity wells, it is simply not the case. Time is not a dimension and gravity is not a force. "So what is gravity?"

We have studied the effects of what gravity does for millennia. Gravity is what we perceive as a force, but it is the by-product of how mass interacts with reality.

H

Hedonism

A philosophical approach that posits pleasure and happiness as the ultimate goals in life. It emphasizes the pursuit of pleasure and the avoidance of pain as the central aspects of human existence.

Hedonism places great importance on individual agency and the fulfillment of personal desires and goals. It suggests that one should actively seek out experiences and actions that bring pleasure and contribute to overall well-being. This perspective recognizes that each individual has the autonomy to determine their own path to happiness and personal fulfillment.

I

Immortality

Refers to the idea of living forever, without experiencing death or the end of life. It suggests that a person or being continues to exist indefinitely, without the natural process of aging or the finality of death. Immortality is often portrayed in various myths, legends, and works of fiction, where characters possess eternal life or the ability to cheat death. In simple terms, immortality means never dying and living on forever.

Information

Knowledge or data that we gather and use to understand things and make decisions. It is the valuable content or facts that help us learn, communicate, and make sense of the world. Information refers to the organization or arrangement of the elements within a system.

"IT From BIT"

A term coined by physicist John Archibald Wheeler, and it means that the core of everything in the physical world comes from information. This concept says that when we ask yes/no questions and get answers from the tools we use, we're essentially creating the reality we see. In simple terms, it means that the universe is like a gigantic computer that processes information to shape our reality.

This idea is connected to the weird rules of quantum mechanics, which say that particles don't have fixed places or speeds until someone looks at them or measures them. This means that our observations and measurements are what make the physical world real. In other words, the universe isn't just made of matter and energy; it's also made of information that's always changing and being processed.

The "IT from BIT" idea has big implications for how we understand the universe and what we think is real. It suggests that information isn't just something extra; it's a fundamental part of the universe that's behind everything we see and experience.

J

Joy

A strong and pleasant feeling of happiness and delight. It is the pure and uplifting emotion that fills our hearts with positivity and brings a sense of inner contentment and fulfillment.

K

Karma

The belief that the actions we take in life have consequences that determine our future experiences. It is the notion that our deeds, whether good or bad, shape our destiny and influence the circumstances we encounter.

L

Levinthal's Paradox

When looked at through the lens of the Conscious Anthropic Principle (CAP), Levinthal's Paradox gives us important insights into how proteins fold, shared information and the purposeful creation of the universe are all connected. Proteins use a shared pool of knowledge and follow shared patterns of information to fold correctly. This lets them fold quickly and accurately without needing to try out countless shapes.

When we bring in the CAP, we see a bigger picture. The CAP says the universe is conscious and it wants intelligent life to develop. This challenges the idea that the universe is just a random mix of particles and energy. Instead, it suggests that the universe is thoughtful and has a purpose. In this view, consciousness and shared patterns of information work together to shape the universe and create the specific conditions we see in reality.

With the CAP in mind, Levinthal's Paradox shows us how consciousness, shared information, and life's basic processes are all connected. It tells us that proteins, influenced by collective consciousness and guided by shared patterns, fold as part of the universe's purposeful creation. Protein folding is like a dance that shows us the universe is conscious and everything is connected.

By starting with the CAP, we get a better understanding of the paradox. We see how collective consciousness, shared information, and conscious intention are all part of making proteins fold correctly. This perspective challenges the idea that everything is random, showing us a deeper purpose and connection in the universe.

M

MBTI

The MBTI is based on the conceptual theory proposed by Swiss psychiatrist Carl Jung, who speculated that people experience the world using four principal psychological functions — sensation, intuition, feeling, and thinking — and that one of these four functions is dominant for a person most of the time. The four categories are introversion/extraversion, sensing/intuition, thinking/feeling, and judging/perceiving. Each person is said to have one preferred quality from each category, producing 16 unique types.

Metaphysics

The branch of philosophy that studies the fundamental nature of reality, the first principles of being, identity and change, space and time, causality, necessity, and possibility.

Morphic Resonance

A concept in biology and philosophy proposed by Rupert Sheldrake. The idea suggests that there is a non-material or energetic field that connects all living things and that this field has a memory of sorts, which can influence the behaviour and development of organisms.

Morphic resonance posits that "memory is inherent in nature" and that "natural systems... inherit a collective memory from all previous things of their kind."

According to Sheldrake, this field, called the morphic field, contains the collective memory and habit patterns of a species, which are transmitted from one generation to the next through a process of resonance. This means that the more a behaviour or pattern is repeated by a species, the more likely it is to be imprinted on the morphic field and become part of the species' collective memory.

The concept of morphic resonance has been controversial and criticized by many scientists for its lack of empirical evidence and scientific rigour. However, some researchers have explored the idea of non-local communication and information transfer in biology, which could provide a possible mechanism for the transmission of information across generations and between living organisms.

Overall, the concept of morphic resonance remains a topic of debate and speculation, and further research is needed to determine its validity and potential implications for our understanding of biology and evolution.

The Multiverse

Multiple universes exist, having all possible combinations of characteristics, and we inevitably find ourselves within a universe that allows us to exist.

N

Nihilism

Nihilism posits that life is inherently devoid of inherent meaning, purpose, or value. According to this perspective, existence is ultimately meaningless, and individuals must create their own subjective meaning.

O

Oddism

A philosophy emphasizing one universe and one shared experience, captured in the present moment—Right Now. It promotes seeking simplicity in understanding the universe's seemingly overwhelming complexity, steering clear of parallel realities. The core idea is that all information for the past, present, and future is encapsulated in the present itself. The past reflects data anchored in the present, while the future projects using this same anchor. This underscores the pursuit of wisdom within the present, emphasizing the important role of the current moment as the linchpin for both past and future.

Optimism

A mental attitude or perspective characterized by a positive outlook and a hopeful belief that things will generally turn out for the best. It is the tendency to focus on and expect favourable outcomes, even in the face of challenges or adversity. Optimistic individuals often possess a sense of confidence, resilience, and a proactive approach to life.

P

Panpsychism

The view that mind or a mindlike aspect is a fundamental and ubiquitous feature of reality. It is also described as a theory that "the mind is a fundamental feature of the world which exists throughout the universe." It is one of the oldest philosophical theories and has been ascribed to philosophers including Thales, Plato, Spinoza, Leibniz, William James, Alfred North Whitehead, Bertrand Russell, and Galen Strawson.

Panpsychism vs CAP
The Conscious Anthropic Principle (CAP) does share some conceptual similarities with panpsychism but represents a distinct perspective on the nature of the universe and consciousness. While panpsychism proposes that consciousness is a fundamental aspect of all matter and exists at some level in everything, CAP focuses on the interaction between collective consciousness and shared fractal information.

CAP suggests that the universe is conscious first and that reality emerges from this conscious foundation. It proposes that the universe is not merely a random collection of particles and energy but rather a purposeful creation that fosters the development of intelligent life. Consciousness, in CAP, interacts with shared fractal information to shape the universe and give rise to the finely tuned conditions we observe.

In essence, CAP posits that the universe is a contemplative creation, intentionally designed to support the emergence of intelligent life. It suggests that conscious intentionality is driving the construction of the universe, utilizing shared patterns of fractal information.

While panpsychism and CAP both explore the relationship between consciousness and the universe, CAP takes a specific perspective that emphasizes the intentional nature of the universe's creation and the interaction between consciousness and shared fractal information. It provides a unique framework for understanding the purposeful development of intelligent life within the universe.

Panspermia (From Ancient Greek Seed)

The hypothesis, first proposed in the 5th century BC by the Greek philosopher Anaxagoras, that life exists throughout the universe, distributed by space dust, meteoroids, asteroids, comets, and planetoids, as well as by spacecraft carrying unintended contamination by microorganisms. Panspermia is a fringe theory with little support amongst mainstream scientists. Critics argue that it does not answer the question of the origin of life but merely places it on another celestial body. It is also criticized because it cannot be tested experimentally.

Panspermia proposes (for example) that microscopic lifeforms which can survive the effects of space (such as extremophiles) can become trapped in debris ejected into space after collisions between planets and small Solar System bodies that harbour life. Panspermia studies concentrate not on how life began, but on methods that may distribute it in the universe.

Pseudo-panspermia (sometimes called soft panspermia or molecular panspermia) is the well-attested hypothesis that many of the prebiotic organic building blocks of life originated in space, became incorporated in the solar nebula from which planets condensed, and were further — and continuously — distributed to planetary surfaces where life then emerged. — Wiki

Panpsychic Life Theory

This theory advocates that all life emerged from shared information and ancestry. This theory also suggests that all extraterrestrial civilizations are currently at a similar stage of advancement.

Pantheism

Pantheism is the belief that the universe and nature itself are divine or sacred. It sees the entire cosmos as an interconnected and conscious entity, where the divine is not separate from the world but is inherent in every aspect of it.

The Participatory Universe

A physics concept suggesting that the act of observation or measurement by a conscious observer plays a vital role in shaping reality. This idea was introduced by physicist John Archibald Wheeler, who proposed that the universe is not solely an objective, physical reality but also a subjective, participatory one, influenced by the observer.

Physics

The natural science that studies matter, its fundamental constituents, its motion and behaviour through space and time, and the related entities of energy and force. Physics is one of the most fundamental scientific disciplines, and its main goal is to understand how the universe behaves.

Q

Qualia

Refers to the subjective and personal experiences we have as conscious beings. It encompasses the raw sensations, perceptions, and feelings that make up our unique human experiences, such as the taste of chocolate, the warmth of sunlight on our skin, or the beauty of a sunset. Qualia highlights the richness and diversity of our individual perspectives in understanding the world around us.

Quantum Mechanics

A branch of physics that deals with the behaviour of particles at the smallest scales, such as atoms and subatomic particles. It has some fundamental principles that help explain these particles' strange and fascinating behaviour of these particles. Here are a few key principles of quantum mechanics:

1. Wave-Particle Duality: Quantum mechanics suggests that particles like electrons and photons can behave both as particles and as waves. This means they can exhibit behaviours of both solid particles and the wavelike patterns you might see in waves on water.
2. Superposition: Particles in quantum mechanics can exist in a superposition of states. This means they can be simultaneously in multiple states until measured or observed, like Schrödinger's famous cat that is alive and dead until you look at it.
3. Quantization: Some particle properties are quantized, meaning they can only take on specific discrete values. For example, the energy levels of electrons around an atom are quantized.
4. Uncertainty Principle: This principle, formulated by Werner Heisenberg, states that there's an inherent limit to how precisely we can know both the position and momentum of a particle at the same time. The more accurately we know one of these values, the less accurately we can know the other.
5. Quantum Entanglement: When two particles become entangled, their properties become linked, even if they are far apart. Changing the state of one particle instantly affects the state of the other, no matter how far apart they are.

6. Measurement Problem: Quantum mechanics raises questions about what happens when we measure or observe a particle. The act of measuring can actually change the state of the particle, making the outcome probabilistic.
7. Quantum States and Operators: Instead of precisely describing the properties of a particle, quantum mechanics uses mathematical entities called wave functions to represent probabilities of different outcomes. Operators are used to perform operations on these wave functions.
8. Quantum Tunneling: Particles can "tunnel" through barriers that classical physics would predict they couldn't cross. This is how some phenomena, like nuclear fusion in stars and electron transport in semiconductors, work.

While often counterintuitive compared to classical physics, these principles have been confirmed through numerous experiments and have led to the development of technologies like lasers, transistors, and more. Quantum mechanics is a cornerstone of modern physics and has revolutionized our understanding of the micro world.

Quantum Suicide

A thought experiment within quantum mechanics and the philosophy of physics. It aims to explore the differences between the Copenhagen interpretation and the many-worlds interpretation of quantum mechanics. The concept is derived from Schrödinger's cat thought experiment and was developed by Max Tegmark.

In this thought experiment, a person stands in front of a gun that is designed to fire based on the detection of a subatomic particle's spin. If the particle is detected to have an upward spin, the gun fires, and if it has a downward spin, the gun does not fire. The process is repeated multiple times. From an external perspective, both outcomes are equally likely. However, in the many-worlds interpretation, it is argued that the person in front of the gun would always find themselves in a universe where the gun never fires, even though this outcome is highly improbable.

This thought experiment suggests that according to the many-worlds interpretation, a person could perceive themselves as immortal since they can only remain conscious in a world where the gun does not fire. However, it is important to note that this is a hypothetical scenario and does not reflect real-world expectations of immortality.

Max Tegmark has clarified that the reasoning behind expecting general immortality is flawed. In reality, the process of dying is gradual and involves a progressive decrease in consciousness rather than an instantaneous event. The thought experiment presents a specific imaginary scenario, and it is within this scenario alone that a person could perceive themselves as surviving.

It is worth noting that most experts believe that the quantum suicide experiment would not actually work in the real world and that the thought experiment serves primarily as a theoretical exploration rather than a practical demonstration of immortality.

R

Religious Existentialism

A philosophical perspective that combines existentialist ideas with religious beliefs and faith. It acknowledges the existentialist notion that human existence is marked by freedom, responsibility, and the search for meaning. However, it adds the dimension of religious faith and the belief in a higher power or divine presence. Religious existentialists believe that through their individual choices and actions, they can find meaning and purpose in alignment with their religious beliefs. They see their relationship with God or the divine as integral to their existence and the source of ultimate meaning. This perspective emphasizes the importance of personal faith, the exploration of existential questions within a religious context, and the integration of religious values into one's existential journey.

S

The Scientific Method

The scientific method is a systematic approach used by scientists (and inquisitive minds) to investigate and understand the natural world. It involves a series of steps that help formulate and test scientific hypotheses and theories. The core steps of the scientific method typically include:

1. Observation: Scientists make careful observations of phenomena or events in the natural world.
2. Question: Based on the observations, scientists formulate a specific question or problem to be addressed.
3. Hypothesis: A testable explanation or prediction is proposed to answer the question. It is typically based on prior knowledge and observations.
4. Experimentation: Scientists design and conduct experiments to test the hypothesis. They manipulate variables, control conditions, and collect data in a controlled and repeatable manner.
5. Data Analysis: The collected data is analyzed using statistical methods and other techniques to draw meaningful conclusions. The analysis helps determine whether the hypothesis is supported or refuted.
6. Conclusion: Scientists interpret the results and draw conclusions based on the analysis. They evaluate the hypothesis and discuss its implications and limitations.
7. Communication: Scientists share their findings through scientific papers, presentations, and discussions. This allows the scientific community to review and replicate the experiments, fostering knowledge exchange, and further scientific progress.

The scientific method is an iterative process, and the conclusions reached often lead to new observations, questions, and hypotheses, initiating the cycle again. This method provides a rigorous and systematic framework for investigating the natural world, promoting objectivity, and advancing scientific knowledge.

Schrödinger's Cat Paradox

A thought experiment in quantum mechanics that questions the concept of superposition. It proposes a scenario where a cat inside a closed box can exist in a state of being both dead and alive simultaneously, challenging our understanding of quantum states. Schrödinger introduced this paradox to critique the Copenhagen interpretation of quantum mechanics.

What's intriguing is that Schrödinger himself did not hold a materialistic view and believed in a universal Mind that explained everything, which suggests a non-materialistic perspective.

Solipsism

Solipsism, in philosophy, is an extreme form of subjective idealism that denies that the human mind has any valid ground for believing in the existence of anything but itself.

Stoicism

A philosophical school of thought that originated in ancient Greece, around the 3rd century BC, and later became popular in ancient Rome. The philosophy emphasizes personal virtue, wisdom, rationality, and self-control as the path to achieving a tranquil and contented life, even in the face of adversity.

Key principles and tenets of Stoicism include:
- Virtue as the Highest Good: Stoicism teaches that the only true good is virtue or moral excellence. Virtue is defined by qualities such as wisdom, courage, justice, and self-discipline. Stoics believe that living a virtuous life is

the key to eudaimonia, which is often translated as "happiness" or "flourishing."
- Acceptance of What Cannot Be Controlled: Stoics emphasize the importance of distinguishing between things we can control (our thoughts, actions, and attitudes) and things we cannot control (external events, the actions of others, and the past). They argue that we should focus on what we can control and accept the rest with equanimity.
- Apathy (Ataraxia) and Emotional Resilience: Stoics advocate for emotional detachment or ataraxia. This doesn't mean suppressing emotions but rather achieving emotional resilience by not being overly attached to or disturbed by external events. Stoics believe that by controlling one's reactions and judgments, one can maintain inner tranquillity.
- The Dichotomy of Control: The Stoic Dichotomy of Control is the idea that everything in life can be divided into two categories: things we have control over (our own thoughts and actions) and things we do not have control over (external events and other people's actions). Stoics focus on the former and accept the latter.
- Living in Accordance with Nature: Stoics believe that living in harmony with nature means living in accordance with reason and the natural order of the universe. This involves understanding and accepting the impermanence of life and the inevitability of change.
- Duty and Social Virtue: Stoicism emphasizes our duty to society and the importance of acting justly and virtuously in our interactions with others. This includes treating others with fairness, kindness, and respect.

Prominent Stoic philosophers include Epictetus, Seneca, and the Roman Emperor Marcus Aurelius, whose writings have been influential in spreading Stoic ideas. Stoicism has had a lasting impact on Western philosophy and continues to be a source of practical wisdom for those seeking to navigate life's challenges with a calm and rational mindset.

Superconductivity

A phenomenon in which certain materials, when cooled below a critical temperature, can conduct electric current with zero resistance. In other words, the electrical resistance drops to virtually nothing, allowing the flow of electricity without any energy loss. Superconductors also exhibit other unique properties, such as the expulsion of magnetic fields (Meissner effect) and the ability to maintain persistent currents. This phenomenon has significant practical applications in various fields, including energy transmission, medical imaging, and quantum computing.

Exciton condensates, on the other hand, refer to a collective state of matter that arises from the coupling of excitons. Excitons are quasiparticles consisting of an electron and a positively charged "hole" (an electron vacancy) in a solid material. When a large number of excitons condense and behave collectively as a single entity, it forms an exciton condensate. Similar to how superconductivity involves the collective behaviour of paired electrons, exciton condensates exhibit coherence and can display unique properties, such as the ability to transport energy efficiently. Exciton condensates have potential applications in the development of new types of optoelectronic devices and fundamental research on quantum phenomena.

T

Teleology

Teleological perspectives argue that life has an inherent purpose or goal towards which it is directed. They propose that the meaning of life lies in fulfilling this predetermined purpose or reaching a certain end.

Time

Can be thought of as a representation of causality, through the eye of a conscious observer.

Time is an integral part of our reality, connected to the expansion of the universe. It is not a separate dimension but rather an emergent construct that emerges from the movement of the present moment to the next moment. Causality, the cause-and-effect relationship between events, exists within this progression. The expansion of the universe is the driving force behind this emergent construct of reality, and all other forces are derived from it.

The core essence of reality is the emergence and expansion of information, resembling a fractal pattern. Conscious thought can only exist in the present moment, while the past exists as information and the future is yet to come into existence. This understanding of time clarifies why there is an arrow of time, the directionality of cause and effect, and why time travel is considered impossible. Time can be seen as a representation of causality, observed through the perspective of a conscious observer.

Time Travel

We seem to have only one lifetime in this form. This lifetime is unique and non-repeating. Certain elements may be cyclical, but the experience is unique. The concept of time travel may exist but in a life cycle framework and is beyond human experience.

Theism

Theistic perspectives find meaning in the belief in a higher power or divine being. They propose that the purpose of life is to fulfill a religious or spiritual destiny set by that higher power.

The Truman Show Delusion

Also known as Truman syndrome, it is a type of delusion in which the person believes that their lives are staged reality shows, or that they are being watched on cameras.

U

Unfalsifiable

Not capable of being proved false. — Webster

About the Author

Meet Todd R LeBlanc, a seasoned lab manager with a remarkable three-decade-long career dedicated to the intricate workings of various space laboratories. Todd's expertise spans across electronics, fabrication, integration, manufacturing, testing and metrology, marking him as a valuable participant and inquisitive observer in the advancement of the commercial space field.

Todd is distinguished by his extensive experience and insatiable curiosity. With an unwavering commitment to understanding the intricacies of the world, Todd possesses a relentless drive to question everything and unravel the mysteries of everyday life. This inquisitive nature leads him to delve deeper, continuously seeking answers to the fundamental "why?" questions in pursuit of knowledge.

Beyond the realms of science, Todd is an awkwardly polite and sensitive observer with a profound interest in austere religion and philosophy, exploring different cultures and belief systems. This passion for embracing varied viewpoints and gaining insights into the human mind adds a unique depth to Todd's work, acknowledging the profound impact that diverse perspectives have on humanity's understanding of existence.

Todd is intellectual with a practical side, embracing tangible aspects of his craft, and bridging the gap between theory and practice.

Described as an "old soul," Todd possesses a unique blend of expertise, animated curiosity, and a dry, sarcastic personality. His wisdom and depth were shaped by "the school of hard knocks" and fueled by an insatiable sense of wanderlust. This background has significantly influenced and deepened Todd's view of the world and helped produce his distinctive contributions to the scientific community.

IN SEARCH OF THE UNKNOWN

The night was bitterly cold, and a sharp wind seemed to change directions with every passing moment, lending an ominous sense of foreshadowing to the unknown adventure ahead. Through the labyrinthine streets of the bustling city's underbelly, the adventurers navigated with determination. Their destination: the Dragon's Tale, a renowned tavern shrouded in dim lighting, known to attract dangerous individuals and seekers of mysteries. Upon entering the warm establishment, the air became infused with the inviting scents of ale and anticipation.

Seated around a sturdy wooden table, the adventurers took turns introducing themselves. The Cleric, a wise and devout follower of a forgotten deity, spoke passionately of their divine purpose and the mission to bring light to a world shrouded in darkness. The Elf, graceful and mysterious, unveiled a profound passion for the art of combat. The Ranger, a skilled tracker, regaled the group with vivid tales of the untamed wilderness and their unwavering commitment to preserving the delicate balance of life. Lastly, the brooding and stealthy thief grunted and made his presence known. Our party's enigmatic aura filled the dark room.

Finally, it was my turn. Clearing my throat, I addressed the group, "Greetings, my esteemed companions. I am Odd, the Sorcerer." All eyes turned toward me, captivated by the enigma standing before them.

Continuing, I declared, "Tonight we start living our dreams. We start simple and build towards the spectacular. This great adventure signifies our maiden voyage into comprehending the depths of our own reality." Pausing for a moment, determination gleamed in my eyes.

The adventurers leaned in, their curiosity piqued by the promise of an extraordinary journey. Inhaling deeply, I shared my knowledge of a long-lost legend — the enigmatic dungeon known as the Caverns of Quasqueton. Rumoured to harbour unimaginable treasures and ancient artifacts, its entrance had remained shrouded in mystery for centuries.

Whispers circulated among the group, hinting that the dungeon's location lay hidden within the pages of a forgotten tome — an artifact concealed within the depths of an abandoned library. The book was said to be fiercely guarded by peculiar creatures, formidable traps, and riddles that would challenge even the most astute minds.

As the flames flickered in the hearth, casting dancing shadows upon the tavern's walls, we solemnly pledged ourselves to embark on a perilous quest — a quest to unveil the mysteries, confront the monstrous inhabitants, and retrieve the legendary book. Bound together by our shared thirst for knowledge and our unwavering hunger for adventure, our party meticulously prepared for the arduous path that awaited us.

This marked the beginning of many things, but it was the first time I dedicated time and effort to understand my reality. It laid the foundation for countless grand quests and provided a time for introspection. I have spent a lifetime searching for answers to profound questions — history, faith, science, and the relentless pursuit of purpose. And so, this is the tale of Odd the Sorcerer, chronicled in the Book of Odd.

---

www.ingramcontent.com/pod-product-compliance
Lightning Source LLC
Chambersburg PA
CBHW031101080526
44587CB00011B/774